Computers in Biomedical Research

VOLUME III

Contributors to this Volume

ROBERT P. ABBOTT

JAMES O. BEAUMONT

DONALD W. CHAAPEL

JEROME R. COX, Jr.

J. R. CUNNINGHAM

E. C. DELAND

M. A. EVENSON

HARRY A. FOZZARD

HARVEY S. FREY

M. M. GIESCHEN

G. P. HICKS

L. A. KAMENTSKY

F. C. LARSON

J. MILAN

WILLIAM J. MUELLER

FLOYD M. NOLLE

G. CHARLES OLIVER

JOHN J. OSBORN

JOHN C. A. RAISON

ALLAN C. SPRAU

RALPH W. STACY

W. NEWLON TAUXE

HOMER R. WARNER

BRUCE D. WAXMAN

Computers in Biomedical Research

Edited by
RALPH W. STACY
COX HEART INSTITUTE
KETTERING, OHIO

BRUCE D. WAXMAN
HEALTH CARE TECHNOLOGY PROGRAM
U. S. PUBLIC HEALTH SERVICE
WASHINGTON, D. C.

VOLUME III

1969

ACADEMIC PRESS   NEW YORK AND LONDON

COPYRIGHT © 1969, BY ACADEMIC PRESS, INC.
ALL RIGHTS RESERVED
NO PART OF THIS BOOK MAY BE REPRODUCED IN ANY FORM,
BY PHOTOSTAT, MICROFILM, RETRIEVAL STYSTEM, OR ANY
OTHER MEANS, WITHOUT WRITTEN PERMISSION FROM
THE PUBLISHERS.

ACADEMIC PRESS, INC.
111 Fifth Avenue, New York, New York 10003

United Kingdom Edition published by
ACADEMIC PRESS, INC. (LONDON) LTD.
Berkeley Square House, London W1X 6BA

LIBRARY OF CONGRESS CATALOGUE CARD NUMBER: 65-15773

PRINTED IN THE UNITED STATES OF AMERICA

List of Contributors

Numbers in parentheses indicate the pages on which the authors' contributions begin.

ROBERT P. ABBOTT, Research Data Facility, Institute of Medical Sciences, Pacific Medical Center, San Francisco, California (207)

JAMES O. BEAUMONT, Department of Cardviovascular Research and Research Data Facility, Institute of Medical Sciences, Pacific Medical Center, San Francisco, California (207)

DONALD W. CHAAPEL, Medical Applications Division, IBM Corporation, Rochester, Minnesota (145)

JEROME R. COX, JR., Biomedical Computer Laboratory, Washington University, St. Louis, Missouri (181)

J. R. CUNNINGHAM, The Ontario Cancer Institute, Toronto, Canada (159)

E. C. DELAND, The Rand Corporation, Santa Monica, California (1)

M. A. EVENSON, Clinical Laboratories and Department of Medicine, University of Wisconsin, Madison, Wisconsin (15)

HARRY A. FOZZARD, Department of Medicine, University of Chicago, Chicago, Illinois (181)

HARVEY S. FREY, Department of Radiology, University of California, Los Angeles, California (87)

M. M. GIESCHEN, Clinical Laboratories and Department of Medicine, University of Wisconsin, Madison, Wisconsin (15)

G. P. HICKS, Clinical Laboratories and Department of Medicine, University of Wisconsin, Madison, Wisconsin (15)

L. A. KAMENTSKY,* IBM Watson Laboratory at Columbia University, New York, New York (107)

F. C. LARSON, Clinical Laboratories and Department of Medicine, University of Wisconsin, Madison, Wisconsin (15)

J. MILAN, The Ontario Cancer Institute, Toronto, Canada (159)

WILLIAM J. MUELLER, State University of New York, Upstate Medical Center, Syracuse, New York (55)

* Present address: Bio/Physics Systems, Inc., Katonah, New York.

FLOYD M. NOLLE, Biomedical Computer Laboratory, Washington University, St. Louis, Missouri (181)

G. CHARLES OLIVER, Department of Medicine, Washington University, St. Louis, Missouri (181)

JOHN J. OSBORN, Department of Cardiovascular Research, Institute of Medical Sciences, Pacific Medical Center, San Francisco, California (207)

JOHN C. A. RAISON, Department of Cardiovascular Research and Research Data Facility, Institute of Medical Sciences, Pacific Medical Center, San Francisco, California (207)

ALLAN C. SPRAU, Applied Mathematics Division, IBM Corporation, Rochester, Minnesota (145)

RALPH W. STACY,‡ Department of Surgery, University of North Carolina, Chapel Hill, North Carolina (1, 253)

W. NEWLON TAUXE, Section of Clinical Pathology, Mayo Clinic and Mayo Foundation, Rochester, Minnesota (145)

HOMER R. WARNER, Latter-Day Saints Hospital, Salt Lake City, Utah (239)

BRUCE D. WAXMAN, Health Care Technology Program, National Center for Health Services Research and Development, Arlington, Virginia (1)

‡ Present address: Cox Heart Institute, Kettering, Ohio.

Foreword

The reports in this volume are further evidence that computer technology as applied to biomedicine has grown to healthy adolescence. As was to be expected, the more impressive advances have been in those biomedical sciences that most readily lend themselves to reductionism. The traditional realms of clinical medicine have posed difficulties which do not reside in computer technology itself. Although patently suitable for superb clinical care, information relating to the patient care process does not readily lend itself to the rigorous treatment required for computer processing.

It is especially timely that we now assess the potentials of computer technology, both for improving health and advancing health services research. Under public and private auspices a new order and level of effort is in prospect. The present total investment of effort and money in health services research remains disproportionately low, inadequate by far, when viewed against the total national expenditures for health services. Systematic research of unprecedented scope needs to be directed to the major problems that beset the new social policies and imperatives in health services. In this effort, computer technology is an invaluable ally. Those who still suspend judgment of its ultimate utility must recognize that unfulfilled expectations of the past are attributable less to the technique than to the conception and direction of its applications.

This cumulative definition of the state of the art embodies productive guiding principles and methods. In it are given a glimpse of the future in which the contributions of computer science will more surely undergird the evolution of a new order of effectiveness and efficiency of health services. For this, the community of health service investigators is indebted to Drs. Waxman and Stacy and their colleagues.

PAUL J. SANAZARO, M. D.
*National Center for Health
Services Research and Development*

Preface

In the few years since research and research efforts began to be channeled into the application of computers and computer technology to biology and medicine, significant strides have been made. We believe that this progress has been accelerated by several factors which bear mention at this point:

(1) The frontiers of biology and medicine were advanced to a level such that the advantages of computer applications were sorely needed; technique development was timely.

(2) Modern methods of research stimulation and support by support agencies have significantly altered the pace of development.

(3) Publications, such as the first two volumes of this series, have permitted dissemination of information regarding applications and techniques which has kept the field growing at a pace faster than would have been expected only a few years ago.

Thus, it is no surprise that even within the short span of a single decade, the field of biomedical computing has been generated, has matured, and is now at the point of specific application to clinical medicine.

The first two volumes of this series were devoted (for the most part) to the basic science of biomedical computing—the applications of computers to the scientific aspects of biology and medicine. We now deem it appropriate to bring forth this third volume, which will have as its emphasis the use of computers in the medical (patient) care environment.

The chapters contained herein do not include new contributions in the field of hospital data processing (i.e., the storage and retrieval of patient records, etc.), but rather concentrate on the technological aspects of clinical computing. For this reason, we find that we pay much attention to two phases of computer technology which in the early years of the field were of secondary importance. These are:

(1) The use of on-line computing techniques, often in real time and coupled with immediate response reporting procedures.

(2) The use of computers which are dedicated to the task at hand and often are not used for any other purpose.

Such developments were in fact predictable even in the early years of the effort, for it is nearly axiomatic that as technology proceeds, it tends toward combined simplification and specialization. In the case at hand, this trend is by no means complete, and it is now apparent that within a very short additional period, the change in the direction will proceed even farther.

It is now anticipated that within a decade, much of the actual patient care in hospitals will have come to depend on the applications of digital computers. It is believed that most of these will be small, dedicated computers; the field is developing in such a way that data base magnitudes are becoming smaller and analytical methods simpler. Smaller computers and simpler methods mean, of course, that the ultimate cost of such modernized patient care will come down, and that good medical care can be provided in a reasonable economic milieu.

The information provided in the chapters in this volume is bound to be of a preliminary nature. We hope and believe, however, that it will provide a nucleus of methods and procedures which will be developed into the simpler, more practical patterns we anticipate.

The total effort involved in preparing the chapters of this volume has been tremendous. It could only be the results of work by such a large and diversely oriented group as the twenty-three contributors hereto. To each of these, we express our appreciation and that of medical science for the work they have put into their writings. We are sure that in the final analysis, this work will not go unrewarded.

RALPH W. STACY
BRUCE D. WAXMAN

August, 1969

Contents

LIST OF CONTRIBUTORS	v
FOREWORD	vii
PREFACE	ix
CONTENTS OF PREVIOUS VOLUMES	xv

INTRODUCTION. Computers and the Delivery of Medical Care

E. C. DeLand, Ralph W. Stacy, and Bruce D. Waxman

I.	Introduction	1
II.	The Computer and the Management of Complexity	4
III.	Extensions of Current Development	6
IV.	Areas for Future Emphasis	10

CHAPTER 1. On-Line Data Acquisition in the Clinical Laboratory

G. P. Hicks, M. A. Evenson, M. M. Gieschen, and F. C. Larson

I.	Introduction	16
II.	The Clinical Laboratory Problem	16
III.	Hardware Implementation	21
IV.	Software Implementation	27
V.	Performance of the On-Line System	34
VI.	Experience with On-Line Data	44
VII.	Critique of System	50
VIII.	Evaluation	51
	References	52

CHAPTER 2. Automated Multiaccess System for Clinical Work

William J. Mueller

I.	Problems Associated with the Automation of Clinical Areas	55
II.	System Implementation	59
III.	System Performance	74
IV.	Current Clinical Experience	79
V.	Critique	79
VI.	Evaluation	82
	Reference	85

CHAPTER 3. An On-Line Graphic Inquiry System for Analysis of Tumor Registry Data

HARVEY S. FREY

I. Introduction 87
II. Implementation 96
III. Performance 102
IV. Critique 103
V. Discussion and Another Solution 104
References 105

CHAPTER 4. Cell Identification and Sorting

L. A. KAMENTSKY

I. Introduction 107
II. Characterization of Cells 109
III. Analysis of Measurements 117
IV. The Rapid Cell Spectrophotometer (RCS) 127
V. Clinical Applications of the RCS 136
VI. Summary 143
References 143

CHAPTER 5. Use of a High-Speed Digital Computer in Processing Radioisotope Scintiscan Matrices

W. NEWLON TAUXE, DONALD W. CHAAPEL, AND ALLAN C. SPRAU

I. Introduction 145
II. Present Technique 146
III. Results 148
IV. Conclusions 158
References 158

CHAPTER 6. Radiation Treatment Planning Using a Display-Oriented Small Computer

J. R. CUNNINGHAM AND J. MILAN

I. Introduction 159
II. General Description 161
III. The Radiotherapy Environment 163

IV.	Data Input	166
V.	Graphic Output	168
VI.	Data Manipulation	169
VII.	Evaluation	171
	References	178

CHAPTER 7. Some Data Transformations Useful in Electrocardiography

Jerome R. Cox, Jr., Harry A. Fozzard, Floyd M. Nolle, and G. Charles Oliver

I.	Introduction	182
II.	Transformations that Eliminate Redundancy	183
III.	Data Storage	186
IV.	A System of Processors for ECG Rhythms	188
V.	Results and Conclusions	200
	Appendix	204
	References	206

CHAPTER 8. Computation for Quantitative On-Line Measurements in an Intensive Care Ward

John J. Osborn, James O. Beaumont, John C. A. Raison, and Robert P. Abbott

I.	Introduction	207
II.	Description of the Computer System	208
III.	Electrical Safety	216
IV.	Respiratory System and Analysis	217
V.	Cardiac Output	223
VI.	Calibrations	224
VII.	ECG Analysis	227
VIII.	Alarms	229
IX.	Exercise Laboratory	233
X.	Evaluation of the System	234
	References	237

CHAPTER 9. Computer-Based Patient Monitoring

Homer R. Warner

I.	Introduction	239
II.	The Monitoring System	241
III.	Summary	249
	References	251

CHAPTER 10. The Comprehensive Patient-Monitoring Concept

RALPH W. STACY

I.	Introduction	253
II.	The Nature of Death	254
III.	Crisis-Prevention Interference	262
IV.	The Real Purpose of Patient Monitoring	263
V.	The Requirements of a Patient-Monitoring System	265
VI.	The State of the Art of Patient Monitoring	268
VII.	A Practical Comprehensive Patient-Monitoring System	271
VIII.	Summary	274
	References	275

AUTHOR INDEX 277

SUBJECT INDEX 280

Contents of Volume I

General Introduction
RALPH W. STACY AND BRUCE D. WAXMAN

SECTION A. Computers and Mathematics on the Life Sciences

1. Complex Dynamic Models of Living Organisms
   LEONARD UHR

2. New Mathematical Methods in the Life Sciences
   GEORGE B. DANTZIG

3. Statistical Packages in Biomedical Computation
   WILFRID J. DIXON

4. The Analog Computer in the Biological Laboratory
   JAMES E. RANDALL

5. Hybrid Computers in Bioscience
   WILLIAM SILER

SECTION B. Computer Simulation of Life Processes

6. Simulation of Biochemical Systems
   DAVID GARFINKEL

7. Computer Simulation of Cybernetic Systems
   FRED S. GRODINS

SECTION C. Computer Analysis of Specific Biosystems

8. Computational Methods in Protein Structure Analysis
   CHARLES L. COULTER

9. The Evolutionary Dynamics of Two-Gene Systems
   KEN-ICHI KOJIMA

SECTION D. Computer Uses in Neurophysiology

10. Computer Analysis in Neurophysiology
    W. R. ADEY

11. A Neural Network Model and the "Slow Potentials" of Electrophysiology
    BELMONT G. FARLEY

12. The Application of Computers to Electroencephalography
    MARY A. B. BRAZIER

SECTION E. Computers in Clinical Medicine

13. Computer Techniques in Medical Diagnosis
    LEE B. LUSTED

14. Computers in Multiphasic Screening
    MORRIS F. COLLEN, LEONARD RUBIN, AND LOUIS DAVIS

15. Computer Assisted Data Processing in Laboratory Medicine
    BALDWIN G. LAMSON

16. Computer Analysis of the Electrocardiogram
    HUBERT V. PIPBERGER

17. Computer Aids in Evaluating Fetal Distress
    EDWARD H. HON

18. Computation of Radiation Dosages
    THEODOR D. STERLING AND HAROLD PERRY

Contents of Volume I

SECTION F. Computation in Psychology and Psychiatry

19. The Computer as a General Purpose Device for the Control of Psychological Experiments
    GEORGE A. MILLER, ALBERT S. BREGMAN, AND DONALD A. NORMAN

20. Computer Simulation of Neurotic Processes
    KENNETH MARK COLBY

21. Automatic Personality Assessment
    HOWARD P. ROME, PETE MATAYA, JOHN S. PEARSON, WENDELL M. SWENSON, AND THOMAS L. BRANNICK

22. Perceptrons as Models of Neural Processes
    J. A. DALY, R. D. JOSEPH, AND D. M. RAMSEY

AUTHOR INDEX—SUBJECT INDEX

Contents of Volume II

SECTION A. Computer Technology in the Life Sciences

    1. Analog-Digital Conversion Systems
        JOSIAH MACY, JR.

    2. A Description of the LINC
        W. A. CLARK AND C. E. MOLNAR

    3. Special Purpose Digital Computers in Biology
        JEROME R. COX, JR.

    4. A Special Purpose Analog Computer for Biochemical Research
        JOSEPH J. HIGGINS

SECTION B. Programming for Life Sciences Research

    5. Systems Programs to Accomodate Biomedical Research
        ROBERT H. BRUNELLE

    6. Programs as Theories of Higher Mental Processes
        ALLEN NEWELL AND HERBERT A. SIMON

    7. Compartmental Analysis in Kinetics
        MONES BERMAN

SECTION C. Biological and Physiological Applications of Computers

    8. Simulation of Ecological Systems
        DAVID GARFINKEL

    9. Computer Simulation of Atrial Fibrillation
        GORDON K. MOE

CONTENTS OF VOLUME II

10. Some Computer Techniques of Value for Study of Circulation
    HOMER R. WARNER

11. Computations of Respiratory Mechanical Parameters
    RALPH W. STACY AND RICHARD M. PETERS

SECTION D. Computers in Hospital Automation

12. Hospital Automation via Computer Time-Sharing
    JORDAN J. BARUCH

SECTION E. Computers in Psychology and Psychiatry

13. Computer Analysis of Psychopharmacological Data
    ROLAND R. BONATO

14. Computer Analysis of Psychological and Psychiatric Data
    HARVEY F. DINGMAN

AUTHOR INDEX—SUBJECT INDEX

Computers in Biomedical Research

VOLUME III

INTRODUCTION

Computers and the Delivery of Medical Care

E. C. DeLAND

THE RAND CORPORATION, SANTA MONICA, CALIFORNIA

RALPH W. STACY

COX HEART INSTITUTE, KETTERING, OHIO

BRUCE D. WAXMAN

HEALTH CARE TECHNOLOGY PROGRAM, NATIONAL CENTER FOR HEALTH SERVICES RESEARCH AND DEVELOPMENT, ARLINGTON, VIRGINIA

I.	Introduction	1
II.	The Computer and the Management of Complexity	4
III.	Extensions of Current Development	6
	A. Problems in Clinical Chemistry	7
	B. Patient Monitoring	7
	C. Problems in Computer Graphics	8
	D. Multiphasic Screening	8
	E. File Structures	9
IV.	Areas for Future Emphasis	10
	A. Problems in Model Building	10
	B. Image Processing	11
	C. Computer-Assisted Instruction	12

I. INTRODUCTION

More than three years has elapsed since the publication of the first two volumes of "Computers in Biomedical Research." In introducing these earlier volumes we observed that the use of computers in biomedical research was still very much in a state of evaluation, and that significant accomplishments were relatively hard to document except in very general terms. We also acknowledged the more deliberate use of computers by biomedical scientists and expressed the hope that in the future these machines would be more definitively productive of substantial biomedical accomplishments.

With the publication of this third volume we have an opportunity to examine these accomplishments over this three-year period. The choice of material that has been included in this volume is in no sense fortuitous. This entire volume has been dedicated to practical applications of computers and, in particular, to their use in the delivery of medical care. While it would have been possible to compile a volume oriented to basic biomedical research, we believe the ultimate test of the impact of the computer in medicine is its ability to assist in the process of the alleviation of morbidity and mortality. The chapters of this volume, therefore, are intended to be illustrative of the degree to which the computer is being effectively utilized in the delivery of medical care.

Despite the heavy emphasis on clinical applications, a reading of this volume cannot help but leave the impression that a definitive answer to the question "Has the computer materially altered the treatment of disease?" is still in abeyance. In some instances, such as in the case of the patient-monitoring activities described in Chapters 7, 8, and 9, there is evidence that lives have been saved as a result of the timely presentation and analysis of vital data. On the other hand, these experiences are not conclusive, either, in the sense that they imply that such monitoring is a panacea for the treatment of the critically ill or that as much could have been accomplished without such elaborate instrumentation.

To complicate matters even further there is the entire issue of whether any increased benefits derived from such instrumentation can stand the test of cost–benefit analysis. It is difficult to ignore the issue of whether a technique that costs tens of thousands of dollars and can be administered only to a limited number of individuals located at a center of medical excellence is truly of significant medical value.

It is in this sense that we contend that definitive answers are not yet at hand. Certainly, as advocates of biomedical computing, we are highly enthusiastic over these accomplishments, but at the same time we are making a valiant effort to preserve our objectivity. Similar observations can be made about the applications of the computer to the problem of radiation dosimetry, Chapter 5, and its use in the on-line monitoring of the ECG, Chapter 6. Note, however, in these latter chapters that an element of economic feasibility is implicit in the developments discussed.

Chapter 3, which deals with the use of a computer to control and analyze data from a very novel cell-counting device, warrants special

Introduction 3

attention for two reasons. First, it is an excellent example of the use of a computer as a process controller, and, in this sense, illustrates the importance of process-control technology in the automation of analytical instruments. Similar examples can be found in Chapter 1, where the automation of clinical chemistry laboratories is described. Chapter 3, however, has value over and above its implication for the use of computers as process controllers in that it also suggests an entirely new technological approach to the highly critical problem of cell evaluation. If this device can swiftly and accurately characterize cell pathology, we have at our disposal an instrument with enormous potential for mass screening. The potential for saving life from undetected cervical cancer alone would more than justify its utilization. Nevertheless, the instrument still must be viewed as very much in a state of evaluation, and a discussion of its use in this volume should not be construed as a testimony to its clinical efficacy.

Chapter 4 has been included in this compendium not only because it heralds significant advances in the clinical use of radioisotope scanning by virtue of the data reduction capabilities of a computer, but also because it represents a practical example of computer-assisted image enhancement. Interest in the processing of biological images has taken on considerable emphasis in recent years, with interests ranging from basic research in artificial intelligence to pattern recognition to the enhancement of radiographs and micrographs. To date there are relatively few examples of established image-processing techniques in an applied area of medical or biological activity. While the discussion of the image-enhancement techniques employed in the radioisotope studies represents a relatively limited and uncomplicated illustration, it has special significance by virtue of its immediate clinical applicability. In this sense it may well be indicative of similar or analogous uses of such techniques in medicine in the relatively near future.

To a certain extent, our enthusiasm for graphic displays and nonpassive data management systems is illustrated rather succinctly by several of the chapters of this volume. While such projects are still a long way from completion, it should be obvious that by virtue of a highly interactive display device supported by a clever file structure and a large library of programs for testing statistical hypothesis the investigator can figuratively "mentally exhaust" himself in a matter of a very few minutes. The prospect of file search and hypothesis evaluation on this short a time scale with pictorial data presentation suggests that the widening gap that now exists between the vast capabilities of the computer to provide data and man's ability to digest the information that is being presented to him may ultimately be eliminated.

II. THE COMPUTER AND THE MANAGEMENT OF COMPLEXITY

To continue the view expressed earlier, a definitive answer to the question "Has the computer materially altered the treatment of disease?" is still in abeyance. However, there is a widespread conviction that this question will be answered affirmatively, perhaps in the very near future. What is the source of this optimism? Is it well founded? In the absence of any preponderance of real evidence and with several dismal failures recorded, will computer systems eventually fulfill the glowing prophecies for their future in medical applications? For the authors of this introduction something more than pure faith is required each day. As consultants and employees of the National Center for Health Services Research and Development, we are constantly required to affirm the sources of optimism and to attempt to assess the important problems, difficulties, and probabilities of the long-range goals. This situation will be discussed in two parts: (1) technological potential of the computer in medical matters and (2) organizational comprehension of the system.

The computer has been dramatically successful in other applications. Few single technological developments in the history of man have had such sweeping and immediate impact on the way we live as the digital computer. It arrived so quickly and so easily became a part of the routine operations of society that often we are not aware of the numerous advantages our complex society derives from it, nor of the somewhat less tangible human freedoms—mobility, efficiency, and flexibility—directly accountable to its use.

We need look no farther than the telephone to note the remarkable increase in flexibility and human freedom because of computer control of the underlying machine. Today we can dial any other telephone in the United States; no doubt soon we will dial any telephone in the world and shortly thereafter any individual in the world with no intervening delay or human operations. Generally speaking, however, these remarkable applications of the computer and the derivative benefits have been in corporate data-management systems and in basic and developmental research, areas in which the motivations for an operating system are very strong. While we in Biology and Medicine understand the needs for the computer and can even perceive its benefits, that same level of motivation to accomplish operating systems does not exist. Typically the instigation for development of such a system in biomedicine comes from the operating personnel rather than from the management.

Although the pervasiveness of computer utility is already difficult to measure and its impact on the functioning of society is clearly as significant as the printing press or the radio, we have experienced (except in rare instances) only routine, not to say mundane, applications of the potential of the computer as a functioning biomedical tool. Computers in medicine have been used primarily to handle tasks that humans, given the time and motivation, could do alone; and, for the most part, this applies to bioscientific research as well as information-handling systems. The advantages the computer has had to offer, then, are not in kind but in quantity. The utility of the computer so far has been principally found in its speed. Speed of operation combined with inhuman precision, memory, and reliability have been the characteristics of the computer primarily responsible for its insertion into the everyday affairs of biomedicine.

However, recently (and in rare instances in the past) other functional qualities of the computer have been acknowledged, especially its complexity. A computer, however complicated, is not just a machine, but a machine plus its attendant software, without which it would be idle. Together they constitute a system that has a history in time, and with each accretion the system becomes more complex. Clearly, this complexity lends a new quality that allows the formulation and organization of tasks that previously could not be contemplated for lack of a sufficiently powerful tool.

A second characteristic distinguishing recent computer systems from those of the past is flexibility; a third is the ability for human operators to communicate with the computer systems at interactive consoles in natural language; and a fourth is increased capacity. In addition to speed, a human operator may now demand of the computer system that it be available essentially whenever he wants it, that it have the capability to accommodate his problem (whether information management or scientific computing), that it "understand" a semiformalized language when he writes it, and within reasonable bounds that it be unrestricted as to size and cost-efficiency.

We are at a point, then, in the development of computer technology when we may begin to take the computer system for granted and begin worrying seriously about medical applications. We may assume systems that can extend, complement, and enhance human problem-solving capability. Software problems remain, but there is every reason to believe that hardware technology will easily keep abreast or remain ahead of our ability or imagination to use it. With such a tool, a human investigator potentially becomes a giant; the limitation of his investigatory range is no longer time or complexity, but his own ability, intuition, and imagination for defining and structuring his problem for computer-aided analysis. This symbiotic relationship between human cognition and the idio-savant computer has opened and will continue to open important new areas of investigation in medicine. It is, perhaps, time to take seriously the possibility of stating and solving the "next generation" medical problems. These problems may be an order of magnitude more difficult than the feasibility studies carried out so far, but they may turn out to have effective biological relevance.

With such tools becoming available, researchers are pressed to formulate clearly the biomedical problems to be solved and to pursue complex problems in their manifold ramifications. The difficulties of designing software systems for analysis of data or for information management resolve themselves eventually in the researcher's understanding of his complex problem. Some problems may be theoretically unsolvable using our current logical concepts, but a great deal remains to be accomplished within the framework of solvable problems. What do we mean by "patient monitoring", which parameters are critical for which disease, and how should the information be analyzed and presented for maximum utility (see Chapter 9)? What should be the information content of a national tumor registry for it to be useful and effective for cancer therapy? How does the thyroid control system function? Are there significant inferential correlations between molecular structure and biological functions?

These difficult questions have two characteristics: (1) they are formulated in biological-science and not computer-science language (i.e., the investigator is not considering whether a computer system can be built to do this task, but is thinking about his problem in biomedical and biomathematical terms), and (2) as with most problems of this nature, there is hope of empirical, experimental, or analytical solution, once one has clear statements of the problem's content and a sufficiently complex tool for its implementation. The significance of this is that piece by piece these problems can probably be solved now. Software systems can be accumulated and, with time and effort, made sufficiently complex to cope with the intricacies of important classes of biological and medical problems.

Obviously, we are optimistic about the role of computer-based technology in the future of medicine. It is difficult to underestimate the usefulness of a technology that appears limited only by our ability to translate biomedical problems into an equivalent or complementary computer system and by the current cost of such systems, which is decreasing rapidly to practical levels. We always bear in mind that classes of problems exist for which no general solutions are known (e.g., the simulation of human cognition); but well inside that theoretical boundary, startlingly complex and useful tasks can be performed.

III. EXTENSIONS OF CURRENT DEVELOPMENT

It is important to delineate several distinct areas of current research and development from which significant contributions to the delivery of

Introduction   7

health services may come. We call particular attention to these areas more to stimulate further development than to convey new information. All of these applications (still in their infancy) have their practical and theoretical traps and bottlenecks. Nevertheless, the following discussion will indicate the status of each case with the logical extensions as it appears now.

A. Problems in Clinical Chemistry

Chapter 1 discusses in detail the problems and accomplishments of automated systems for clinical chemistry. Natural extensions of these data systems involve the incorporation of clinical data into the automated records of the patient's daily chart, his permanent clinical record, and the accounting records of the hospital. It is only a matter of scale, then, to consider central national data files. Aside from the possible ethical questions involved, it seems inevitable that increasing federal and state health services will lead to automated central file systems.

Undoubtedly, model building and mathematical analysis of an individual patient's clinical chemistry data will yield much more patient information than is currently extracted. Patterns of fluid and electrolyte distribution and the response of the mathematical model to chemical stress are evidently indicative of conditions in the whole body for a variety of circumstances. Since early in the history of fluid- and electrolyte-balance studies there has been a conjecture in the literature that blood is a biopsy of the body. With more powerful analytical tools available through the computer and in the clinical laboratory, this conjecture may be demonstrated for a significant class of abnormalities. A much closer working relationship between the clinic and the laboratory is in prospect.

B. Patient Monitoring

The two essential reasons for developing automated patient-monitoring systems are the following: (1) As the relative shortage of medical and paramedical personnel grows with the hospital population, more automation of nursing and paramedical duties is needed. (2) A sophisticated, automatic system can gather more and better information because all critical parameters can be monitored continuously and the data can be presented in analyzed, correlated, and contextual form.

Feasibility studies for patient monitoring applied to particular syndromes (e.g., cardiology and shock) have been described in Chapters 7, 8, and 9. These studies show that it is possible to design automated systems for comprehensive patient monitoring. However, much work concerning the analytic characterization of disease states remains to be done. Shock, for example, a very difficult state to define, depends upon a variety of parameters, some unknown. It is highly time-dependent, as the critical parameters can vary greatly from moment to moment. We need a complex model of the system that will correlate analytically the signs and symptoms with parameters that can be measured and effected. Chapter 10 discusses the problem in some depth and suggests a new role for patient monitoring.

If this level of sophistication is possible, then in addition to merely monitoring, systems can be designed to materially aid the judgmental tasks of the physician. The data display could, in principle, simultaneously present a contextual model of the current state of the patient along with norms and standards and correlations with patients of similar type, and suggest therapies drawn from analyses of banks of patient data and predictions of future outcome.

C. Problems in Computer Graphics

Computer graphics (see Chapter 3) deals with augmenting the problem-solving capability of the human by giving him an easily accessible, flexible tool with inhuman capacity for memory, speed, arithmetic, and logic—a tool with which he can easily build and test conceptual models of his hypotheses concerning complex systems. This problem does not directly concern the substitution of clever programs for human capabilities, but rather a functional symbiosis of the two, each doing what it does best. The potential is great, and it is now sufficiently clear that highly interactive computer terminals are a prerequisite to the symbiosis that a great deal of effort is currently being devoted to the development of interactive graphics. Such terminals and the attendant central-processor software will be useful for hospital communications, for model building and hypothesis testing, for patient autointerview, for perusing files of data, and for other tasks requiring the transfer or analysis of data.

D. Multiphasic Screening

Chapter 4 deals particularly with the problem of screening wide segments of the population for detection of disease. The classic example is the fleet of x-ray mobiles distributed to neighborhood areas for screening the population for TB. The goals are broader now—namely, to screen for a wide variety of diseases. While one important disease (e.g., a class of heart diseases) would be sufficient to justify a program of health screening, routine tests for detection of many diseases are already available. The greatest future tasks will be in expanding the class of screening tests and bringing the force of modern technology to bear on automating and analyzing complex tasks.

Considering the mobility of the national population, the cost and relative immobility of clinical equipment, and the tendency to centralize health care in clinics and hospitals, screening activities for preventive medicine and detection of disease should be located at fixed centers. This augers well for automation, since such equipment usually has high data-rate communication lines to a central processor and large central files.

E. File Structures

Chapter 3 illustrates dramatically a way in which a large computer with capabilities for intensive man/machine interaction via graphic media becomes a sophisticated tool for perusal, analysis, and interpretation of large data bases. The marriage of high quality, interactive graphics with a powerful central processor and massive, direct-access storage results in a system that extends a human's data-analysis efforts and provides him with a sophisticated repertoire of data-transformation procedures. With these he may derive information given only implicitly in the file. While much research and development remains to be done in large scale system design, high-level computer languages, and graphic input/output devices to exploit this concept of information handling to the fullest, a good approximation to the requisite computer technology is already available. The most serious problems are related to representing and organizing biomedical data inside computing machines.

Much of the utility of a data-management system such as that described in Chapter 3 stems from a careful choice of categories into which the data base is organized. If a given query has been "anticipated" a priori in the development of the classification schema, the performance of this data-management tool should approach the responsiveness and precision of an airlines reservation system. If a given query is essentially "unanticipated," however, serious problems result, even though the pertinent data resides explicitly in the file. At best, these data are extracted by a clever concatenation of the basic search and retrieval procedures provided by the file-system designer. At worst, the data are obtained only if one is prepared to write an ad hoc retrieval program that performs a serial search upon the entire data base. In either case, there is a clear-cut requirement for file-organization methodologies leading to freer classification schemes and placing lighter demands on human ingenuity and persistence.

One attactive approach to the development of such file-organization methodologies is based upon the list-processing capabilities of modern computers. The concept, oversimplified for emphasis, is to construct files in the form of such complex data structures as trees or graphs, rather than as a series of individual records. In this manner complex data-item interrelationships not only can be portrayed, manipulated, and searched, but can serve as the basis for simple deductive procedures. These complex data structures in information systems could represent an important step beyond the "unit record" ideas characterizing and limiting many current efforts.

Despite the power and promise of list-processing capabilities, the limiting factor in the development of sophisticated search and retrieval systems for biomedical research is their data base. The human partner must be able to articulate for the machine the particular data structure to be employed, whether it is a mathematical model or a graph interrelating a set of parameters. The data structure is an attempt at formalization of one's current knowledge in a particular area. Consequently, one's ability to know and to know how he knows determines the completeness and the utility of the data structure. The explication and use of these data structures should provide challenging research problems for those interested in the codification of biomedical knowledge, and the results of this research should find important practical application in biomedical information retrieval.

IV. AREAS FOR FUTURE EMPHASIS

In addition to the developmental areas described above, there are others having clinical or research significance not covered in this volume. Three important areas in which some work has been done and in which the potential for contribution to future medical knowledge and for the delivery of health-care services is particularly large are (1) model building, (2) image processing, and (3) computer-aided instruction. Each of these presents extraordinarily difficult problems of comprehension and application; each involves (in fact is dependent for development upon) complex computer systems; and each involves an unusual amount of multidisciplinary research. In return, it is probably true that these areas are the best cost-effective investments available in biosciences data-processing research today.

A. Problems in Model Building

In a functional sense, every abstract conception of a real biological system is a model (e.g., a verbal description or a set of chemical reactions) insofar as it aids the human to perceive relationships or mechanisms in the system. However, we ordinarily think of a computer-related model as an aid for analytic description of the behavior of the system—a statistical representation, a set of differential equations, or a "heuristic" simulation of human cognition. These behavioristic models have been generated for

Introduction 11

many years with varying degrees of success. In some cases, the analytic description may not have depth of insight or biological significance. Now, however, it is possible to design and (with the computer) to build nontrivial mathematical models of complex biological systems. Some of these models are sufficiently complex and the simulations sufficiently valid that medically useful results will come from their further development. Indeed, mathematical computer models of the artificial kidney, models for drug therapy, models of blood chemistry, and many others now have potential clinical application.

In the future, mathematical models combined with graphic and flexible forms of data presentation could become an integral part of research hospitals, monitored patient wards and clinics, and teaching environments, to say nothing of basic research laboratories. The literature is beginning to discuss the possibilities of a "mathematical man" for experimental purposes, diagnosis, teaching, and trial therapy. Although such a development will require considerable basic research, many of the intermediate goals will also have practical application.

In the broader sense, model building may be equated with hypothesis and conjecture formation concerning the design and function of a biological subsystem. Insofar as such a model can be translated into analytic or statistical terms, computer systems can generally be designed to test them. To this end there exist, for example, generalized biomedical statistical packages, generalized chemical-system simulators, and generalized neural-network simulators. There remain the tasks of accumulating and validating an increasing repertoire of such tools and of making them easily accessible not only at the computer but for transfer to other installations. A level of tool generation and transferability for the problems of the physical sciences exists that has not yet been achieved in the biological sciences, owing principally both to the sheer lack of knowledge of biological mechanisms and to the lack of an abstract symbolic language for the representation of biological systems. Both areas are appropriate fields for further research related to information processing.

B. Image Processing

The automated processing of biological images for the purposes of pattern recognition will occupy an increasing amount of computer power and analytic effort in the foreseeable future. There are two essential branches of this research (image enhancement and pattern recognition) which involve a spectrum of difficulty, from filtering noise to the simulation of human cognition. The purpose, however, is always to extract and process information from an image, and to represent that information at a precognitive level for a decision processor. The decision processor may be human, but in some important instances need not be.

Aside from the simulation of human cognition (which is not addressed here), there are generic classes of difficult problems in this field. But, again, the potential payoff is very high. The problems involved include such particularly refractory and complex tasks as acquisition of valid images, the identification of medically significant parameters, interpretation and analysis of results in a medically oriented context, and extension of the analysis with patient-management problems.

An "image" may be an ECG, an x ray, a thermogram, or a scan signal of a microscope slide. More abstractly, an image may be merely a detectable pattern of data in a tumor registry. Generally, but not always, in order to justify design of a system, the images would occur in large numbers, as in mass screening of subsets of the population. However, with increasing reliability and sophistication of systems, it is conceivable that they could be used for guiding diagnoses by nonspecialist physicians.

ECG analysis can now be handled via a telephone line to a central computer, where the automatic analysis is fail-safe in the sense that with acceptable probability all normals are rejected while suspected abnormals are referred, along with a tentative diagnosis, for human judgment. Similarly, Papanicolaou smears could be analyzed, or cell and tissue typing could be accomplished, or simple patterns of signs and symptoms could be extracted from patient-autointerview data.

The utility of automated systems for such applications depends heavily in each case upon the development of clever software for the fail-safe enhancement and analysis of complex information sets. While generically such problems may be classed as tasks in "image processing," many of them skirt dangerously close to very difficult or unsolved problems in pattern recognition, automated inference, and context analysis. Such problems may be inherently more interesting, but they confuse the issue when we intend to address problems for which an explicit solution can now be written down by a sequence of algorithms—in particular, problems concerning the delivery of health-care services.

C. Computer-Assisted Instruction

Four principal areas in the biosciences around which discussion has centered for application of computer-assisted instruction (CAI) are:

(1) Medical school curricula;
(2) Continuing medical education (postgraduate and practicing physician);

(3) Lay education (preventive medicine);
(4) Extramedical education (bioengineering and research).

Of these, only the first has received wide attention. Not all feasibility studies have been equally successful, but it now seems clear that a trend for the future is the use of CAI for augmenting and in some cases replacing (possibly reduced) formal curricula.

Complex systems have been designed in which a student may simultaneously dissect an organ, observe an expert dissection on film, refer to references in the tape library, listen to a lecture on the dissection, and observe closeup slides of each operation. Again, a computer-based system can now lead a student through a complex sequence of probable diagnoses in a short time, presenting tests, checks, and learning devices as required.

A commercial market exists, but first a great deal of careful thought must be spent on the course work and upon the use of this potentially sophisticated tool in nontrivial ways; to date this has just begun.

CONCLUSIONS: What have we learned over the last three years? At the risk of being doctrinaire, we the editors are of the opinion that:

(1) The heightened interest in the more applied uses of computers through the delivery of medical care has naturally had a much greater impact on the practice of medicine than the use of computers in basic biomedical research, though in the long run the latter will be more profound. What seems to have happened is that the use of computers in clinical applications has had a catalytic effect in bringing basic life-science research to bear more directly on the practice of medicine. Biophysics and physiology no longer necessarily have to remain remote from the patient.

(2) Perhaps the single most important contribution that computer science has made to medical practice is in the application of process-control techniques. This form of technology, more than any other, has been responsible for tangible progress to date, largely because it has made possible the direct application of computers to laboratory and patient-care environments.

(3) The implications of the previous observations have considerable relevance as to the type of computers that are most useful in on-line biomedical investigation. Without disparaging or failing to acknowledge the importance of the very large computer configuration, we believe that it is nevertheless apparent that there is a prior need to attach smaller machines to the patient, to analytical instruments, to experimental animals, etc. This need in on-line biomedical research and, in particular, medicine would seem to attach secondary importance to large computing systems.

(4) The only notable exception to the applicability of the smaller process-control computer is the potential need for graphic displays supported by extensive data-management systems. In this instance, the existence of large computers appears to be indispensable. On the other hand, the need for these capabilities tends to arise in connection with very specific objects rather than as a general-purpose capability lying around awaiting use.

(5) Some of the "grand" experiments in computer science have, to date, been of only marginal value in life-science research. In particular, the emphasis on time sharing over the last three years has, as yet, not had the impact on science in general or bioscience in particular that was predicted. Perhaps these systems will prove to be of more value in the long run and the disappointment has only to do with unrealistic time schedules. It is not obvious to these writers that time sharing is the most fruitful direction to be pursued by those who are interested in getting on with their biomedical research. Once again, this sort of remark has to be tempered by the realization that certain classes of problems, mainly those requiring multiple access to large files, probably do require such capability. However, to the extent that much of biomedical research and application is associated with transducers that produce very high data rates, the argument of economy of scale may not be as salient as that which espouses a multiplicity of small dedicated machines, possibly networked together to perform a hierarchy of functions.

CHAPTER 1

On-Line Data Acquisition in the Clinical Laboratory

G. P. HICKS, M. A. EVENSON, M. M. GIESCHEN, and F. C. LARSON

CLINICAL LABORATORIES AND DEPARTMENT OF MEDICINE, UNIVERSITY OF WISCONSIN, MADISON, WISCONSIN

I.	Introduction	16
II.	The Clinical Laboratory Problem	16
	A. Growth of Laboratory's Services	16
	B. The Impact of Automation	17
	C. Handling Specimens and Reports	18
	D. Application of Computers	19
III.	Hardware Implementation	21
	A. Requirements	21
	B. LINC Computer	22
	C. Interfaces with Laboratory Equipment	24
IV.	Software Implementation	27
	A. System Description	27
	B. Laboratory Monitor Design	28
	C. On-Line Software Design	30
	D. User-Control Programs	32
V.	Performance of the On-Line System	34
	A. Operation by Laboratory Personnel	34
	B. On-Line Reports	37
	C. Specialized Maintenance Programs	40
	D. Non-Automated Functions	42
VI.	Experience with On-Line Data	44
	A. Results	44
	B. Real-Time Quality Control	45
	C. Capacity of System	48
VII.	Critique of System	50
VIII.	Evaluation	51
	References	52

I. INTRODUCTION

There has been much discussion about the development of "computerized information systems" in medicine, especially in hospitals. Many proposals for computer-controlled systems for collation and distribution of all information within hospitals (total hospital information systems) have been made. It now appears that computerization in this area is not as simple a matter as first conceived. There is as yet no complete system known to be in daily operation (Williams, 1967). Some degree of success has been achieved, however, by investigators working on various aspects of the problem.

Because of the nature of the data produced and the suitability of the laboratory operation to computer processing, the clinical laboratory is often the first section, outside of the business office, to develop a computer system. This chapter deals with one such application using the LINC computer. The primary objective of the LINC system is to utilize the computer as a laboratory tool so that data acquisition and processing is a by-product of a normally operating laboratory. There are two major considerations in accomplishing this objective. First, the laboratory personnel should be able to use the computer without specialized training and knowledge of computer technology. Second, the computer should not restrict or force modification of laboratory methodology.

The LINC has proved particularly well suited to this sort of operation. Its capacity to operate in the "conversational mode" makes it easy to use. The success of the conversational mode of operation has been demonstrated by its use in the direct collection of medical histories from patients (Slack *et al.*, 1966; Slack and Van Cura, 1968 a, b) and physical-examination data from physicians (Slack *et al.*, 1967). By applying the same principles the computer can be used by all laboratory personnel. Through the use of a specialized set of "user-control programs" that have been developed, the laboratory personnel control the computer operation and implement changes in laboratory methodology without a detailed knowledge of programming. This laboratory computer system is referred to as LABCOM (*Laboratory Aided By COMputer*).

This chapter presents the overall laboratory problem and describes how the LINC computer has been utilized toward its solution.

II. THE CLINICAL LABORATORY PROBLEM

A. Growth of Laboratory's Services

During the past two or three decades, clinical laboratories have experienced a dramatic increase in demands for their services. Since this trend is a direct outgrowth of the ever-increasing knowledge of the nature of disease, it is likely that it will continue. This growth has been particularly impressive in chemistry, where it is common for a laboratory to increase its work load at the rate of ten to twelve percent per patient per year. The gaining popularity of screening tests has yet to have its major impact and will probably add substantially to the laboratory load.

B. The Impact of Automation

Initially the capacity of the laboratory to respond to the increased work load was limited by the size of the technical staff who performed the procedures manually. This limitation is being removed rapidly through the use of instruments to automatically perform laboratory test procedures.

Of the many laboratory procedures, those in chemistry have been automated to the greatest extent. The procedures first selected for automation were those that are commonly done in large numbers (glucose, urea, electrolytes, etc.) Even complex procedures, such as the assay of serum for protein-bound iodine, have been successfully automated. Procedures that require the recognition of morphologic patterns (e.g., white-cell differential counts, bacterial-colony identification) have proved to be difficult to automate. Although some headway is being made through entirely new approaches to these measurements, automation to the extent that it has been achieved in the clinical chemistry laboratory still seems remote.

With the automation of laboratory instruments, more procedures can be performed with greater precision at lower cost than ever before. It has become practical to include many more standards and controls than before, increasing the accuracy and precision, hence improving consistency from one laboratory to another. On the other hand, substantial changes in staffing patterns have been necessary. Most large laboratories find themselves in the midst of this adjustment. Previously, when procedures were performed manually, the main body of the laboratory staff consisted of broadly trained technical people, usually medical technologists, who were able to do many different procedures. Although laboratory work that requires this level of competence still remains, automation has created a demand for two other levels of competence, one less and the other better trained than the usual medical technologist. The more highly trained personnel require a greater understanding of the electronics and mechanics of often times highly complicated and novel instruments. The lesser trained personnel can perform the more routine tasks of labeling specimens, splitting and processing samples, and even using automated equipment with supervision when it is in good operating condition.

C. Handling Specimens and Reports

Automated instruments have shortened the time required for the actual analysis, but new rate-limiting steps in the total procedure have appeared. Some of these steps can be successfully dealt with through the use of computers.

In order to understand the complexity of the problems of using laboratory services, it is necessary to review the steps involved.

The organization of the clinical laboratory and its relationship to the overall procedures of practicing medicine in the hospital may vary considerably in detail from one institution to another. However, most laboratory systems share the common problems of clerical tasks of ordering and reporting test procedures and data acquisition within the laboratory. The general data-handling problems of the clinical laboratory have been discussed by Lamson (1965), Williams (1964), and Giesler and Williams (1963).

The description of almost any laboratory operation is adequate to present the overall problem. At the University of Wisconsin, for example, the physician enters the laboratory request into an order book separate from the patient's record. The order book, which also contains orders for other services in addition to those performed in the clinical laboratory (e.g. diet, drugs, x rays, etc.), is periodically reviewed by the nursing staff. The nurse collates the routine laboratory test orders. Later a third person, often an attendant or ward clerk, fills out order forms for the tests and enters on the forms the information used to identify and locate the patient, perhaps by the use of an embossed plastic plate and imprinter. The most common specimen sent to the laboratory today is venous blood. Blood may be collected by the attending physician, house staff physicians, medical students, a medical technologist, or nurses. The specimens are carried to the clinical laboratory by an attendant or laboratory personnel. On receipt in the general laboratory area, the specimens often must be subdivided to permit distribution to separate laboratory divisions. This requires proliferation of forms, usually without the advantages of the embossed identification plate for transferring the information. The results of the analysis are entered by the technologist on the order form. One copy of the report is sent to the ward, one is retained in the laboratory files, and a third is sent to the business office for billing purposes. The delivery of the test result can also be a complex procedure. The results from the various divisions of the laboratory may be collated in the laboratory for each patient or several separate reports delivered to the ward. In any case, the report is usually copied into the patient's record by hand, resulting in a cumbersome, difficult-to-read record.

Orders for routine studies are written throughout the day, and most specimens are not collected until the following morning. Thus it is usual for the elapsed time between the physician entering the order and the collection of the specimen to vary from twelve to twenty-four hours. The period of time required for the performance of the analysis and the return of the test result to the ward is usually less than eight hours. The result is ordinarily available for entry into the patient's record on the afternoon of the day following the initial request. Longer delays are experienced over weekends and holidays. Thus the processing of the order, collection of the specimen, and delivery of the result to the ward constitute a major delay.

D. Application of Computers

Several efforts using standard electronic data processing equipment have been initiated to use computers in the clinical laboratory (Lindberg, 1965; Peacock *et al.*, 1965; Straumfjord *et al.*, 1967; Schoop *et al.*, 1968). Applications of computers to laboratory operations can be considered in two categories: (1) data acquisition—the collection of data directly from laboratory instruments (on-line operation) and from laboratory personnel through the conversational mode, and (2) the use of computers to perform essentially clerical tasks. The latter may include the processing of physicians' requests, filing laboratory test results, and communication of the final report to the physician. On-line computer systems for the clinical laboratory have been described (Cotlove, 1965), the first routine on-line application probably being that of Blaivas (1965). Small computers have also been emphasized in some systems (Hicks *et al.*, 1966; Brecher and Wattenburg, 1968; Pribor *et al.*, 1968a). Still others have emphasized the acquisition of data from instruments in computer-processable form with subsequent processing by a computer off-line (Flynn *et al.*, 1966; Rappaport *et al.*, 1968).

Once the various ways in which computers can be applied to the laboratory operation have been identified, the decision whether the laboratory should be totally computerized at once or whether it can be done stepwise must be made.

At the University of Wisconsin a decision was made to introduce computers into the laboratory gradually, in steps that would produce the least possible perturbation of the existing system. The "holistic" approach presented many practical problems that made it unacceptable to us. First, it was expensive to initiate, and, from cost studies, expensive to operate when finally installed. Second, it has been assumed that all is known that is needed to be known about the hardware requirements and that this equipment is available. Third, some systems under consideration have required the use of not only computers but card-punching machines, card sorters, collators, and other ancillary equipment, all of which require additional personnel. This introduces additional clerical personnel into the laboratory who have little understanding of the laboratory operation and who are in short supply. Finally, the software for a total laboratory computer system has not yet been developed. The general problem of selecting a computer system has been candidly discussed by Williams (1967).

In using a stepwise approach to the introduction of computers into the laboratory, the implementation of the overall system has been considered as four separate operations.

(1) Processing the physician's order, collection of the specimen, and delivery of the specimen to the laboratory.

(2) Acquisition of the laboratory test result from the laboratory personnel and/or instrumentation.

(3) Collation of patient identification information, including location in the hospital or clinic, with analysis result (the "patient file").

(4) Communication of final test results to the physician and entry of the results into the patient's chart.

The computerization of these operations can be implemented in a succession of phases. For several reasons, it is convenient to begin with the second operation, the acquisition of data. While the patient file is the ultimate objective, the acquisition of the laboratory test results has a major impact on the structure and quality of the file system. A highly automated patient file system utilizing a manual data-acquisition system, permitting errors to be made by nonlaboratory personnel, can only result in a poor quality patient file. Second, in the acquisition process within the laboratory (especially in real-time, on-line systems) there is a virtual marriage between computer and laboratory methodology. In the laboratory, therefore, the computer has the greatest opportunity to influence the quality of test results. By utilizing the high speed of electronic computers in the acquisition process to provide immediate services to the laboratory in the form of calculations and automatic quality control it is possible to increase both the quality and quantity of test results produced. Finally, the on-line acquisition of test results in the laboratory is one of the simplest of all procedures to run in parallel. To use the computer to perform all of the calculations from automated equipment in the chemistry laboratory requires only the appropriate wires and a switch on the instrument. When the on-line system is operational, it can provide a significant service to the laboratory. On the other hand, certain procedures, such as maintaining patient files, present a much higher risk if there is a failure during the initial phases of becoming acquainted with the new technology of computers. Upon the completion of the data-acquisition system, the collection of patient identification and establishment of the patient file, as well as processing of the physician's orders and communication of the test result to the physician will complete the laboratory information system.

The acquisition of data can be conveniently considered as two types of operations, automated on-line and nonautomated, depending upon whether the instrument output is fed directly into the computer or stored in an intermediate device (including a technologist) for subsequent entry into the computer. A data-acquisition system must include the direct collection of information from both instruments and laboratory personnel.

III. HARDWARE IMPLEMENTATION

A. Requirements

Information in the clinical laboratory is generated by two sources, instruments and personnel. Any computer hardware used in the laboratory must be capable of interfacing with both sources of data. Most automated and semiautomated instruments in laboratories can generate analog signals as their output. The conversion of these analog signals to digital values permits direct entry of the test results into a digital computer.

The interface with laboratory personnel is more complex. In order to collect data directly from all laboratory personnel without requiring a knowledge of computer technology it is necessary for the computer to communicate directly with the personnel in their own language and terminology. This communication can be accomplished through the "conversational mode" of operation with a number of devices such as teletypes and cathode-ray screens using keyboards. With these devices and the appropriate programs in the computer, laboratory personnel can enter data directly into the computer while the computer maintains control in the overall data-acquisition process.

In addition to this interface capability the computer should have adequate mass storage to construct files of laboratory data. The files should hold a minimum of one day's test results with appropriate patient identification and perhaps as much as one month's results from the laboratory for each patient. In any case, since the small computer is limited in its capacity to handle mass storage and retrieval systems, the laboratory computer should be flexible enough to interface with larger central computer systems or networks.

Finally, in order that the system can be utilized by many hospital laboratories, it should be inexpensive and reliable. The use and justification of a computer on the same basis as other laboratory instruments will greatly enhance the acceptance of this new tool.

B. LINC Computer

In 1964, when the University of Wisconsin Clinical Laboratories was searching for an adequate laboratory computer system, there was only one computer available that could meet all the requirements at a reasonable cost. The development of the LINC (*L*aboratory *IN*strument *C*omputer) by Clark and Molnar (1964) with the support of the National Institutes of Health represented a major breakthrough in computer hardware for laboratory applications. A picture of a commercially available[1] LINC is shown in Fig. 1-1. Since its original conception, the LINC has been considerably expanded by industry and made commercially available. In its present form, a typical LINC configuration includes 32 analog channels with analog-to-digital conversion and multiplexer. Eight of the analog channels may be connected to knobs on the front panel of the instrument for direct control of programs, while the other 24 channels can be used for the direct input of signals from various instruments. The precision of the analog-to-digital conversion is 0.2% of full scale. The memory size is 4000 12-bit words. Up to 28,000 additional words of memory can be added. Commercially available LINC computers execute one machine program instruction in 1 to 4 μsec. As auxilliary storage, the LINC uses magnetic tape with the capacity of 131,000 12-bit words per tape. A typical configuration includes four tapes giving about 500,000 12-bit words or 1 million 6-bit characters of storage. A unique feature of the LINC magnetic tape system is that the tapes are premarked and randomly accessible. For high-speed mass storage, a 2-million word disc is available as an option for LINC computers.

The LINC is designed around the cathode-ray screen and keyboard as the standard mode of input and output. The applications of the LINC for computer interviewing using the scope in the conversational mode have been reviewed (Slack and Van Cura, 1968a). In addition, multiple teletype terminals can be added to the computer for both input and output functions. The cathode-ray screen provides the capability of graphic display

[1] The LINC is manufactured by two companies: Digital Equipment Corporation, Maynard, Massachusetts (LINC-8) and Spear, Inc., Waltham, Massachusetts (micro-LINC). While the designs are different in many respects, both machines are based on the "LINC concept" and have been used in clinical laboratory applications.

Fig. 1-1. LINC-8 computer system.

within the laboratory and high-speed communication with personnel where required. By adding multiple teletypes to the computer, many "remote stations" can be implemented throughout the laboratory to provide a slow but inexpensive and effective direct communication with personnel.

In terms of memory speed and size, auxilliary storage, and the ability to interface with instruments and personnel, the LINC is well-suited to the clinical laboratory system. In terms of cost, even an expanded LINC system is very often less expensive than other equipment commonly used in clinical laboratories. The electronic reliability of the LINC has surpassed that of other automated instruments in the laboratory.

A very powerful, but more subtle, advantage of the LINC computer is that it has been designed as an integrated instrument, as opposed to a computer with add-on peripherals. For example, the analog-to-digital conversion hardware with multiplexer and the cathode-ray screen communication system are all part of the standard LINC configuration, and no additional equipment is required. Consequently, the set of instructions and operations used to program the computer is greatly simplified for laboratory operations. For example, to display a character on the screen requires only a single instruction in the computer, and the complex operations of displaying the pattern are handled entirely by the hardware.

Another major consideration is the LINC computer assembly language that is now available. A very powerful conversational LINC assembly program (LAP) has been designed for the LINC by Clark (1967). Because the entire assembler is designed around conversational communication through the cathode-ray screen with program manuscript on magnetic-tape files, programming of the LINC is very rapid. A program change that might require hours on a computer using paper tape for the input of programs can be accomplished in minutes on the LINC. For example, the complete set of programs including all English-language comments for the entire clinical laboratory system is stored on only three LINC magnetic tapes. Any program can be called into the memory and modified within one or two minutes when required. The entire laboratory system can also be readily duplicated in minutes by copying the magnetic tapes.

Finally, the ability of the LINC to interface with a variety of devices permits the communication of information to larger computer systems. As options, the LINC can write standard formated IBM tapes, punch paper tape, and drive card punches and high-speed printers. At the University of Wisconsin, a LINC has been satellited directly to a CDC 3600 computer for the very-high-speed transmission of information directly from LINC core to the large computer.

C. Interfaces with Laboratory Equipment

Most instruments in the laboratory can be made to produce an analog signal as output. Figure 1-2 is a representation of the most common scheme for automating test procedures and interfacing them with computers. In

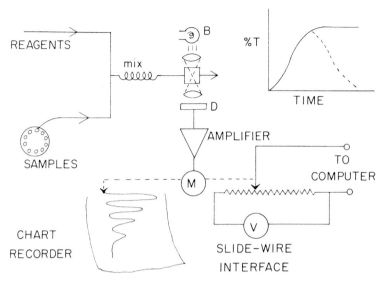

FIG. 1-2. Most common scheme for automation and the computer interface.

its simplest form, the performance of a test procedure involves adding a measured amount of an appropriate chemical reagent to a sample (blood, serum, urine, etc.) and mixing them in a test tube. Upon mixing, a color may increase or decrease in intensity. The magnitude of the color change is proportional to the substance (sugar, proteins, etc.) being measured in the sample. The color change can be quantitated in a colorimeter that measures the percent of light transmitted (%T) through the tube containing the reaction mixture. The %T can in turn be related to the concentration of the sought-for substance from a working curve made by plotting the %T measurements for a series of standard samples having known concentrations. All of these steps—measurement of samples and reagents, mixing, measurement of %T, and calculation of the concentration from the working curve—can be performed manually or automatically.[2]

The AutoAnalyzer[R] method of automating this same procedure is illustrated in Fig. 1-2. The required chemical reagents are mixed with the samples in a flowing stream in an exact proportion. The reaction mixture then flows in a stream through a colorimeter cell. The measurement of the %T in the cell is made continuously by passing a light beam from the energy source (B) through the cell to the photodetector (D). The signal

[2] Automated analysis equipment based on flowing-stream principles is commercially available from the Technicon Corporation, Tarrytown, New York, under the trade name of AutoAnalyzer.

from the detector is amplified and used to drive a motor (M) to produce a continuous ink trace on chart-recorder paper.

Samples are introduced into the flowing-stream system at rates from 20 to 120/hr, depending upon the particular test. Air and/or water are usually introduced between samples, producing a series of "peaks" on the chart recorder. Each peak represents a sample, with the peak height proportional to the concentration of the sought-for substance. A plot of $\%T$ and time is shown in the upper right-hand corner of Fig. 1-2. The solid line shows the trace obtained when a single sample is introduced continuously. The $\%T$ changes until a "steady-state plateau" is obtained. The broken line shows how returning to baseline after each sample produces a "peak" instead of a plateau.

Air bubbles may be used to segment the flowing streams in such automated systems to reduce the amount of "interaction" and mixing of consecutive samples in the stream. The characteristics of flowing systems have been studied by Thiers *et al.* (1967). The general subject of continuous automated analysis and applications has been reviewed by Blaedel and Laessig (1966) and Blaedel and Hicks (1964).

A method of transmitting a form of the output signal from the automated equipment to the computer permits a direct on-line interface. One way to accomplish this is also shown in Fig. 1-2. A voltage (V) is applied across a slide-wire potentiometer. The shaft of the potentiometer is mechanically linked to the same motor that drives the recorder pen. The voltage across one side of the potentiometer is wired directly to a LINC analog channel, usually 0–2 V full scale. In this manner the LINC "sees" an analog signal that is an exact reproduction of the recorder chart as seen by the instrument operator.

In all cases, the interfaces with instruments in the laboratory have consisted of a retransmitting slide wire from the chart recorder on the instrument to the LINC computer. A voltage for the slide wire may be provided by either a battery or a central power supply at the computer. In addition to its ultimate simplicity, there are several reasons for selecting a retransmitting slide wire for the computer interface. In the first place, the retransmitting slide wire is a very inexpensive amplifier when the recorder is already a part of the analyzer. By applying a large voltage across the retransmitting slide wire, a 10-mV signal can be effectively amplified. Second, the mechanical system provides a very effective "damping" of inherently noisy signals. Both of these features have proved highly desirable in the laboratory interface. Finally, and probably most important, by using the transmitting slide wire the computer sees the same signal traces that the technologist sees on her chart recorder. Since even in the case of automated instruments the computer must communicate with the operator, it is essential that the operator and the computer observe the very same signal points. As will be shown later, the computer can in effect reproduce the chart recording on its cathode-ray screen for inspection by laboratory personnel. This permits maximum communication and understanding between the computer and the personnel.

While most of the automated instruments produce analog signals in the laboratory, some measurements are inherently digital. For example, the counting of the number of discrete cells in a sample is a digital measurement. Because of the few digital instruments in use and the probable increase in this type of interface, the on-line computer must accept both digital and analog information. The LINC is well suited for both applications. Wherever an analog measurement is involved, however, the direct transmission of an analog voltage to the LINC without any prior analog-to-digital conversion is the preferred approach and by far the simplest and least expensive.

IV. SOFTWARE IMPLEMENTATION

A. System Description

Figure 1-3 is a diagram of the overall laboratory data-acquisition system. This figure shows the relationship between all the instruments in the

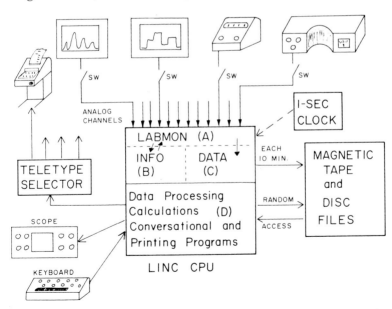

FIG. 1-3. Computer-based data-acquisition system.

laboratory, the mass storage of the computer, and communications within the laboratory in the form of teletypes and the cathode-ray screen with keyboard.

A variety of instruments are connected to the LINC through the analog channels. Each instrument, as shown in Fig. 1-3, has a switch that presents the analog signal directly to the computer when closed. It is this switch that the technologist closes to start a run on-line.

The printed output from the computer, including test results from the on-line system, quality-control comments, and the results for calculations are directed to any one or all of four remote teletype stations through the teletype selector. This permits the on-line results to be printed at the site of the instrument as soon as they are available. Communication with the computer for the purpose of entering data and operating the computer system is done primarily through the conversational mode using the scope and keyboard terminal at the computer. For communications not requiring the scope, data can be entered from the teletype terminals.

Information can be stored in the central processor of the computer or retrieved at any time on random-access magnetic tape and disc files. With this configuration, on-line monitoring of instruments can operate simultaneously with other functions such as communicating with laboratory personnel or performing calculations of various kinds.

B. Laboratory Monitor Design

The design and implementation of the laboratory monitor program (LABMON) is the key to the on-line data acquisition system. It is the monitor system that reads all the analog channels on-line and keeps track of all voltage readings for all channels while the remaining portion of the central processor in the LINC is being used by other programs. The relationship of the monitor program to the overall acquisition system is illustrated in Fig. 1-3. The memory of the computer is divided into four distinct parts, A, B, C, and D. Area A of the memory contains the laboratory monitor program (LABMON), which reads all the analog channels in the on-line system and generates diagnostic information from the voltage readings. Area B holds all information in the on-line system, including which channels have been scheduled on-line, how many readings to take on each channel, how often to take readings, and which areas on the magnetic tape are to be used for data. This area is also used as temporary storage to hold the points on each peak prior to the selection of the peak value for a sample. All bookkeeping information telling about the on-line system is kept in area B. Final values for each peak for each channel in the on-line system are stored in the data area C. Areas A, B, and C together occupy three of the eight available sections in a 2000-word LINC memory. The remaining five sections of memory, section D, can be used by any other LINC programs. For example, section D is used for the programs that process on-line data, perform calculations, print reports in the on-line system, and operate in the conversational mode with laboratory personnel.

Functionally, the laboratory monitor system in areas A, B, and C is always resident in the memory of the computer. Area D holds whatever program happens to be running at a particular time. An external clock provides a pulse once each second, which causes the computer program in area D to be interrupted. The hardware interrupt stops whatever program is running in section D and immediately starts executing the LABMON program in section A. LABMON examines the information in section B, reads the appropriate analog channels, determines the final peak values and diagnostic reading patterns, and stores these data in area C. Thus all analog channels in the on-line system communicate directly with the LABMON program. The 1-sec clock is synchronized with power-line frequency so that the computer is always in synchrony with instruments in the laboratory regardless of any slight variations in line frequency. At the end of each 10-min period or whenever the data area accumulates more than 100 peak values, all data in area C are stored on tape and area C is reinitialized to receive more data. When LABMON saves any information on magnetic tape, the appropriate indicator flags are set in section B. Upon completing the reading of all channels and all appropriate manipulations, the LABMON program automatically returns to whatever program was running in section D when it was interrupted. The time required for the LABMON program to perform all its functions is less than 10 msec out of each second, thereby using only 1% or less of the computer's time. The remaining 99% of the time is used by the programs in section D. Because of the very short duration of the interrupt and the operation of LABMON, a technologist using the scope on the computer is not aware that the on-line system is running. Thus, while the technologist is performing a manual calculation using the LINC, the LINC is simultaneously collecting data once each second from all the analog channels.

Certain programs in section D automatically check the information flags set by LABMON in section B and initiate the data-processing programs for the on-line system whenever LABMON has saved data from area C. The processing programs in area D sort the information saved by LABMON, make all final calculations, and print results through the remote teletype stations. In a LINC memory that has only 2000 words there is seldom room to have more than one kind of program at one time in section D. Consequently, if a program in section D is processing on-line data and printing results from the on-line system, a technologist cannot simultaneously use the computer keyboard. With larger memories available on LINCs, it is now possible to expand programs to permit technologists to use the machine while on-line reports are being generated simultaneously.

To understand the operation of the on-line system, it is important to separate in one's mind the area of memory in the computer that contains the laboratory monitor system from area D in the computer, which can run any LINC program. These areas of the CPU are divided by the solid line in Fig. 1-3. Programs in section D are the same as any LINC program. All manipulations concerning the logistics of the interrupt and the reinitialization of the program in section D each second are handled entirely by LABMON. Thus a program in section D does not have to be aware of the fact that another program is also resident in the same memory. The simplicity of the LINC interrupt has provided a means of greatly extending the application of the computer in the laboratory.

C. On-Line Software Design

The software or computer programs that process the on-line data collected by LABMON have been designed to emulate the actions of a technologist or chemist in the laboratory. Thus, just as the technologist can divide procedures into several modules of taking readings from an instrument, organizing readings on a work sheet, performing calculations, and recording final results, so are the programs in the on-line system constructed. Figure 1-4 illustrates a typical pass through a series of programs in the on-line system to process data for an analog channel. At the top of Fig. 1-4 the processing programs first fetch the last data stored on tape by LABMON. All of the readings for a particular channel are then sorted into that channel's storage area on the magnetic tape or disc. By examining the work schedule entered into the computer by the technologist through the user-control programs, the processing programs determine whether enough information has been collected to make an on-line report. If enough data are available, a calibration curve is constructed and results are calculated for the channel. These functions, as indicated by the solid boxes in Fig. 1-4, are fixed operations that occur for all readings in the on-line system. Further operations on the data, however, are varied and optional. For example, sometimes it is necessary to correct for the drift of an instrument by running the same sample in two different locations on the automatic turntable. One may also wish to correct for sample interaction—the effect of one sample upon the sample that follows it. Interaction correction is particularly important in flowing systems. For

1. On-Line Data Acquisition

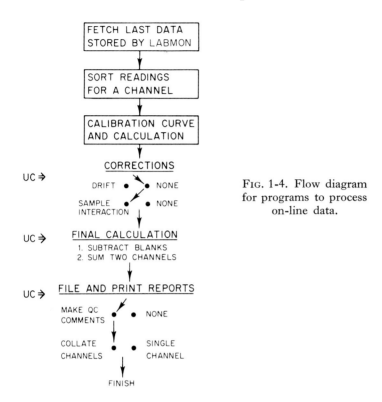

Fig. 1-4. Flow diagram for programs to process on-line data.

complete flexibility, computer programs to perform any one or all of the varied operations upon the data must be available. On the other hand, each analog channel may be unique in the combination of operations necessary for its data. The manner in which this flexibility is attained, where each channel or instrument can be treated separately, is further illustrated in Fig. 1-4. As each section or "module" of the program is used, the appropriate variety of functions is selected. In the particular example shown, the correction-program pathway does not correct for drift but does correct for sample interaction.

After any corrections have been made, additional calculations from the results obtained from the calibration curve may also be necessary. For example, it is sometimes necessary to subtract a blank from the results obtained. Other times the results of two or more instrument channels must be combined to compute the final results desired for the patient's report. At other times the results for corresponding cups on two instruments may be summed to obtain final values.

Finally, all results, including any additional calculated results, must be stored on magnetic tape or disc. Again, it may be desirable to have the computer print certain quality-control comments or not, depending upon the test. One may also desire to have the results from more than one channel printed in a single report, collated cup by cup. For example, it is desirable to print all of the electrolytes (sodium, potassium, CO_2, and chloride) as a single report, since these results are usually considered together. All of these options can be selected as shown in Fig. 1-4.

By designing all of the software in a modular fashion a variety of program options can be selected for each individual instrument that is on-line. The on-line system can be viewed as a pass through a series of programs with the pathway selected by a series of "switches." One instrument may require corrections while the other does not, and so on. A high degree of flexibility is permitted, and the methodology can readily be changed or altered in the laboratory without rewriting programs.

Because of the modular design, new programs that are developed, such as new kinds of corrections or new types of calculations, can be readily inserted into the on-line processing system and selected at the option of the user. Thus expansion of the on-line system is a simple and orderly process not requiring constant redesign.

D. User-Control Programs

The modular design of the programs illustrated by Fig. 1-4 permits the user to exert control over the on-line system at points indicated by UC in the figure. Consequently, a major development that exceeds that of the on-line system itself has been the user-control program system. The user-control system is a set of programs that communicate with the laboratory personnel in the conversational mode. The technologist, for example, enters information about a test procedure or instrument and the programs set up the appropriate program pathways and parameters for that channel. Through the use of user-control programs, laboratory personnel can implement new instruments on-line and change existing on-line procedures through the conversational mode without a knowledge or understanding of computer programming.

The on-line system has been designed to emulate the technologist. By this design, when given all of the information used by the technologist to perform a procedure, the computer can also perform that procedure. For the sake of definition and simplicity, all information used by the technologist has been divided into two categories, *control* information and *schedule* information. Control information is used by the technologist to make judgments and decisions about her tasks. Thus the number of points that should be taken on each peak from an instrument, the acceptable noise level of the signal, the acceptable limits for serum pool controls, and other information used by the technologist as she performs a procedure is stored on magnetic tape for each instrument in the on-line system as a control block. This block also contains the information needed to determine the pathway through the correction, calculation, and printing modules of programs previously discussed. The control block is literally a block of tape containing all of the control information for a particular test.

Schedule information helps the technologist organize her tasks. Thus a worksheet showing which cup in an instrument turntable has standards samples, and controls, as well as telling at what point in the run to make calculations and record results, is considered to be schedule information. This schedule information, which describes the procedure or the order in which the elements of the task are performed, is stored in schedule blocks for each instrument on-line.

By entering new schedule and control information for a test procedure and interfacing the associated instrument to an analog channel of the LINC, new procedures can be readily added to the on-line system. Some typical displays from the user-control programs are shown in Fig. 1-5.

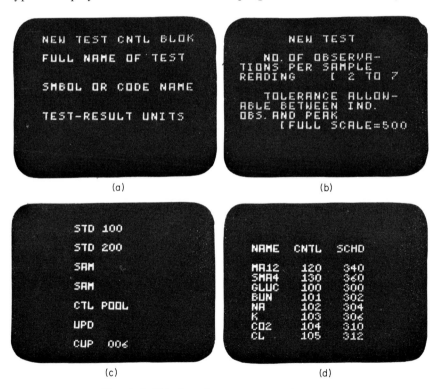

FIG. 1-5. Displays from user-control system.

Display (a) shows how the technologist can enter the full name, code name, and test results units to be used for a new procedure on-line. The new information appears on the screen as she types it on the keyboard. Display (b) requests the number of observations (from 2 to 7) to be made on each individual instrument peak and the acceptable scatter or noise level for readings on the peaks. Displays (a) and (b) are typical of control information.

Display (c) shows some entries for a schedule in the on-line system. Shown in the display are five cups including two standards, two samples, and a pool control. Schedule codes such as STD for standard and SAM for sample are entered sequentially in the same order in which they will be run on a sample turntable. An update code UPD is entered in the schedule whenever a report is wanted from the on-line system. This schedule describes the procedure used for the instrument and is entered only once unless the procedure is changed. This same schedule is used each time the instrument is run on-line. As more schedule codes are entered in display (c), they appear at the bottom of the list, with six codes always being displayed at one time. The cup number at the bottom of the display tells the operator which cup to enter next. Codes on the screen can be deleted by pushing the delete key. Typing in END terminates the display, and the new schedule, if acceptable to both the computer and the laboratory person, is stored in the on-line system for the test indicated.

Display (d) is another user control program that lists the tests currently in the on-line system and the locations of their control and schedule blocks on magnetic tape. Through the user-control programs new procedures and instruments are implemented on-line, deleted, or changed readily without requiring any computer programming. Once an instrument has been interfaced to the analog channels, a new procedure can be implemented for that instrument in about 15 min.

As new programs are added to the on-line system to perform new operations, the selection options are placed into the user-control system. Thus each development of the on-line system itself must be matched by a program in the user-control system permitting the user to operate the program.

V. PERFORMANCE OF THE ON-LINE SYSTEM

A. Operation by Laboratory Personnel

The conversational mode operation of the LINC using either the scope and keyboard or teletype permit the computer to be used by all laboratory personnel without a knowledge of computer programming. The conversational mode is the key to economic operation of the computer without requiring skilled personnel to be added to the laboratory staff. A typical set of displays for operation of the on-line system by the technologist is shown in Fig. 1-6. Display (a) appears on the computer screen at any time

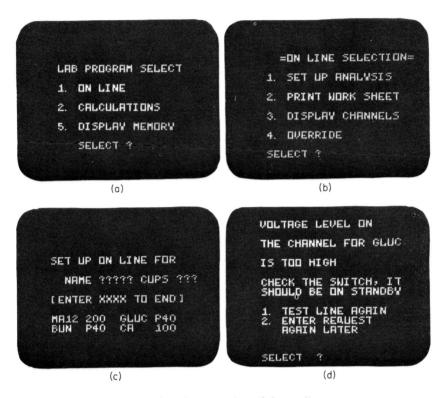

FIG. 1-6. Displays for operation of the on-line system.

the computer is available for general use. The laboratory selection display lists the categories of programs generally available. The computer is operated by selecting various options. For example, pushing the number 1 key on the keyboard and depressing the control key when display (a) is on the screen causes display (b) to appear. Display (b) shows the selection of on-line programs. If a technologist desired to set up an analysis on-line with a particular instrument, she would select 1, in which case display (c) would appear next. Display (c) accepts requests to set up instruments on-line. The same code name used in the control block is entered along with the number of cups to be run. The computer automatically checks the legality of each request, verifying such information as the code name, whether or not the analog channel is available, and even whether the request has already been made by someone else. Display (d) in Fig. 1-6 shows one of the computer checks made during an on-line request by the technologist. This display appears whenever the switch on an instrument has been left on. The instrument switch must be in the off position in order for the technologist to set up the analysis on-line, the computer refusing to accept the request until the switch is again turned off. This kind of checking procedure illustrates how the computer must deal simultaneously with both the instrument and the human interface in order to ensure reliable operation.

When each request is accepted, the name of the test and the number of cups appear at the bottom of the screen in a list and the question marks reappear for the name and number of cups for another test. When the list of tests has been completed, X's are entered for the name. The computer then sets up all of the tests on-line and prints the acknowledgment of each request as it is set up on line as shown in Fig. 1-7.

```
SCHEDULE COMPLETE FOR UREA
40 CUPS SET UP, WILL SCHEDULE
PLATE BY PLATE
COMPUTER ON STANDBY



SCHEDULE COMPLETE FOR GLUCOSE
40 CUPS SET UP, WILL SCHEDULE
PLATE BY PLATE
COMPUTER ON STANDBY


THE FOLLOWING TESTS WILL BE
COLLATED INTO A SINGLE REPORT

GLUCOSE
UREA
```

FIG. 1-7. Computer printouts acknowledging each test set up on-line.

As shown in Fig. 1-6c, two kinds of cups may be requested, simply the number of cups that are to be run or a special code, P40. If, for example, 200 cups are requested for an instrument, the computer simply assumes that when the switch is turned on 200 cups will be run in sequence on the turntable. For some instruments, analyses are performed one plate at a time. For this kind of operation, a request of P40 may be entered. When the switch is turned on, the computer will collect the peak values for 40 cups and then automatically stop. When the switch is turned off for 1 sec and turned back on again, the computer will monitor another 40 cups and stop. Technologists may continue to perform analyses in the "plate by plate" mode for up to 8 plates throughout the day without entering another request.

After entering the on-line request, the technologist returns to the instrument and performs the analysis. When the first peak appears on the chart recorder, the technologist raises the switch to start the computer. The computer then monitors all peaks for the instrument in whatever mode is requested and prints periodic reports on the teletype at the site of the instrument.

B. On-Line Reports

Figure 1-8 is an example of a report printed for the four automated instruments that perform the four electrolyte analyses, sodium, potassium, chloride, and bicarbonate. Reports can be printed for each channel individually or as a collated report for a group of instruments with similar schedules. Figure 1-8 is a collated report. Figure 1-13 is an example of a single report, discussed later.

Collated reports consist of two parts, quality-control comments and the report of results. Comments, which always precede the actual report, may include peak-reading diagnostic patterns (discussed later), a list of standards that have drifted beyond acceptable limits, standards that appear out of a preset order, and any controls that fall outside predetermined ranges. The report lists (in columns) the cup numbers and the corresponding results for each channel. For calibration standards, voltages instead of concentrations are printed, since the technologist already knows the standard concentrations. An asterisk (*) is printed after any result or voltage reading that has a deviant diagnostic pattern. The $<$ and $>$ symbols are used to indicate results that are respectively below or above the range of the standard curve.

The primary purpose of collating data from several instruments is to improve organization and reduce printing time. The comments are based entirely on the control and schedule information entered by the technologist, and considerable flexibility is permitted. For example, in Fig. 1-8 there are reading diagnostics for three channels as shown by the * symbols after the results, but pattern comments were printed only for sodium. This is because the technologist was able to specify that the diagnostic pattern should be printed for each cup only when there are more than five deviants occurring in one printout. This saves unnecessary printing time, since the presence of a few diagnostics may not be meaningful. In any case, all of the comments and results are based on the original control and

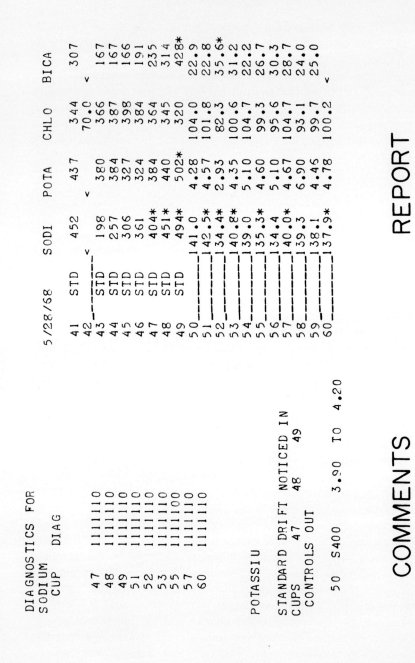

Fig. 1-8. Collated on-line report.

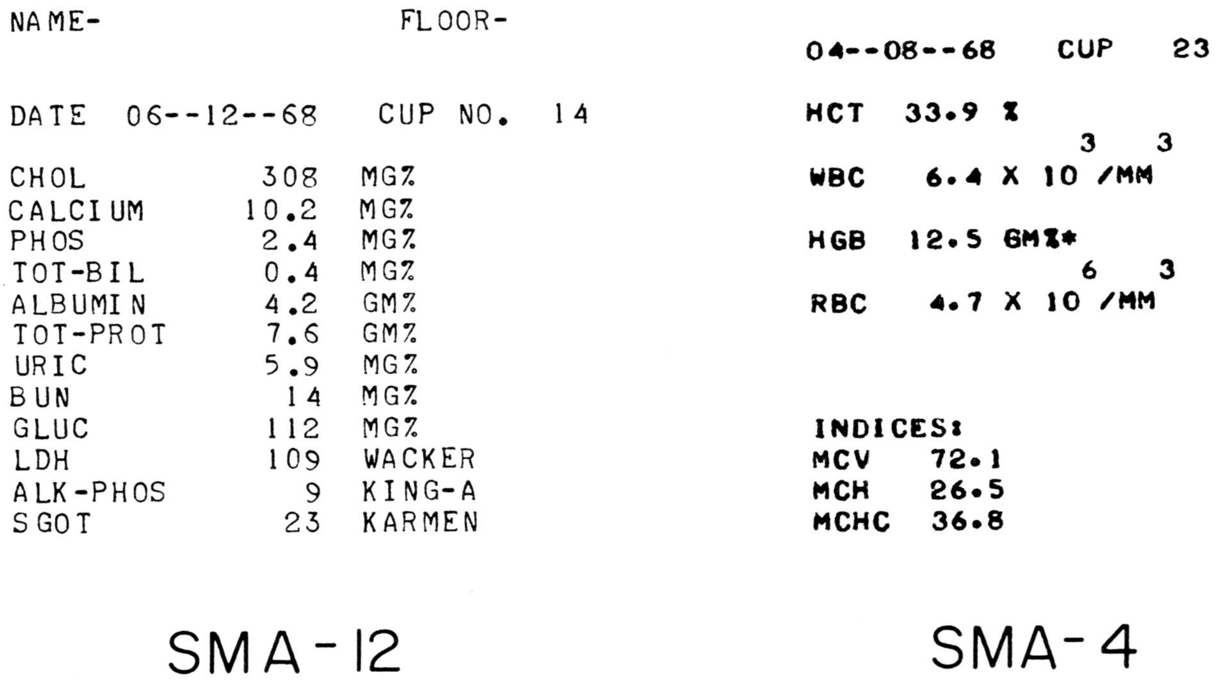

NAME—

FLOOR—

DATE 06--12--68   CUP NO. 14

CHOL	308	MGZ
CALCIUM	10.2	MGZ
PHOS	2.4	MGZ
TOT-BIL	0.4	MGZ
ALBUMIN	4.2	GMZ
TOT-PROT	7.6	GMZ
URIC	5.9	MGZ
BUN	14	MGZ
GLUC	112	MGZ
LDH	109	WACKER
ALK-PHOS	9	KING-A
SGOT	23	KARMEN

SMA-12

04--08--68   CUP 23

HCT 33.9 %

WBC 6.4 X 10³ /MM³

HGB 12.5 GM%*

RBC 4.7 X 10⁶ /MM³

INDICES:
MCV 72.1
MCH 26.5
MCHC 36.8

SMA-4

Fig. 1-9. Multiple analysis reports.

schedule information and are used by the technologist to judge the validity of the computer data.

Recently, automated instruments that split each sample into many parts and perform multiple analyses on each specimen have been introduced into the laboratory. An instrument of this type may perform 12 different analyses on each sample, processing 60 specimens/hr or, expressing it differently, producing 720 results/hr. The output of these instruments is usually traced on a single-pen chart recorder with the output of each colorimeter appearing for only a few seconds. The interface to the LINC for this instrument required only one channel and is essentially the same as a single-channel instrument except that there are 12 results for each cup when the data are processed.

Figure 1-9 shows printouts for the two major types of multiple-analysis equipment in wide spread use. The SMA-12 equipment[3] performs 12 chemical analyses on each sample. The SMA-4 performs four hematology tests on each whole-blood specimen. The index values are computed by the LINC from these four test results.

C. Specialized Maintenance Programs

In addition to the programs that permit the technologist to operate the system, specialized programs to assist laboratory personnel in determining the source of trouble (when problems occur and results from the on-line system are not valid) are also necessary. Such problems may occur as a result of reagents in the instrument, instrument failure, or failures within the computer system. In an interactive system operating in real time the complexity and urgency of the problem present a challenge to the user. A major problem is presented when the computer is treated as a "black box" into which instrument readings disappear and from which results emerge. If there is no provision for laboratory personnel and programmers to inspect the contents of the computer and its storage system during on-line operation, the computer cannot be eliminated as a source of trouble. Through the use of specialized maintenance programs the contents of the computer can be browsed and the analog system can be checked at any time to determine whether the computer, its peripheral equipment, or the laboratory instrument is a source of trouble.

A frequent concern is whether or not the analog system is functioning properly. Figure 1-10 shows two kinds of display that can be used to check the analog system and the operation of the laboratory monitor

---

[3] SMA-12 (Sequential Multiple Analyzer) and SMA-4 are registered trade names of the Technicon Corporation, Tarrytown, New York, for instruments that perform 12 chemical tests and four hematology tests on each specimen respectively.

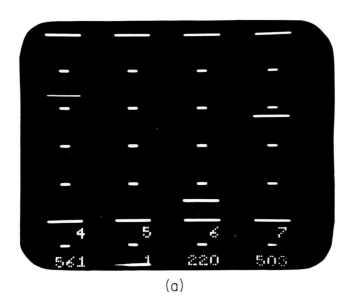

(a)

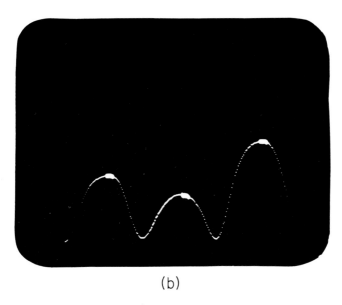

(b)

FIG. 1-10. Displays associated with specialized maintenance programs.

program. Display (a) is a program that plots the voltages as a function of time from four analog channels simultaneously. At the bottom of each channel is the analog channel number (channels 4–7 in Fig. 1-9a) and the voltage reading for the channel. The voltage readings at the bottom of the display are replaced once each second, i.e., each time LABMON takes readings. The display can be used to observe ac noise imposed on the analog lines or to detect problems with the instrument interface. For example, when a retransmitting slide wire becomes dirty or worn out, the voltage level for the channel in display (a) appears broken and noisy. Most problems in the analog system or the instrument interface can be detected in real time.

A second kind of display is shown in Fig. 1-10b. On request, the computer traces the voltage on the cathode-ray screen for any analog channel. The trace on the screen is identical to the trace on the chart recorder for the instrument being monitored. In this manner, the LINC reproduces exactly what the technologist sees on the chart recorder. Whenever the LABMON program takes readings from the analog channel the point on the trace is enlarged. This permits the technologist to determine whether or not the computer is reading peaks from her instrument at the right time. Figure 1-10b is a display of a channel trace that is synchronized properly. In this particular case, seven readings were taken at 1-sec intervals. The peaks appear to be wider on the cathode-ray screen than on the recorder. This occurs because the chart-recorder paper speed on the instrument is very slow and makes peaks appear very sharp when, in fact, if plotted second by second, the peaks are quite broad. In most cases the computer can take seven consecutive readings at 1-sec intervals on a peak with a variation of less than 3 parts in 500.

In addition to the display of voltages on the analog channels, it is often desirable to display the contents of the computer's memory and its mass storage system. By selecting option 5 in the lab selection display in Fig. 1-6a, the contents of memory can be displayed. Information can be read from the magnetic tape and examined while on-line. The ability to examine memory in real time while the instruments are on-line is invaluable for the correction and maintenance of programs. When a dozen or more instruments are operating on-line and only one is having problems, the ability to display the memory and examine the tape is often the only method of obtaining information about the failure. Shutting off the computer would result in the loss of information on all channels.

D. Non-Automated Functions

During on-line monitoring and periods when no reports are being printed for the on-line results, the LINC may be used for various non-automated functions. One such application is the use of the computer as

1. *On-Line Data Acquisition*

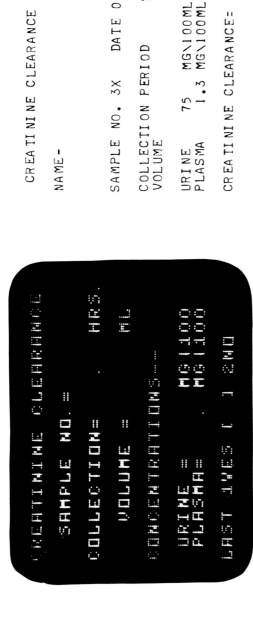

Fig. 1-11. Calculation of creatinine clearance.

a "desk calculator." Frequently, manual calculations must be performed on test results to obtain final results for the patient's record. The data used in the calculation are often produced manually and may not be conveniently interfaced directly with the computer. However, by interfacing directly with the technologist performing the calculation the results can be collected directly by the computer while providing a service directly to the laboratory thereby saving technologist time and improving the accuracy of results.

Typical applications are the calculations of electrophoresis, creatinine clearance, steroids, and iron-binding capacity. An example of the procedure for calculation of creatinine clearances is shown in Fig. 1-11. The sample accession number, urine-collection period, urine volume, and creatinine concentrations for urine and plasma are entered into the display for each specimen. After each set of data the technologist indicates whether it is the last specimen by entering 1 for YES or 2 for NO. Many calculations can be entered at one time. After the last sample, the computer immediately calculates the results and prints the reports as shown. The name of the patient and floor location are written on the report by hand and the printout returned to the patient's record.

The decimal points for the collection period and plasma concentration are already in the display in Fig. 1-11. As the numbers are typed in, the digits enter from right to left over the decimal point, always showing the proper decimal-point location. This technique eliminates confusion over whether "24" means 24 or 2.4, for example.

Performing calculations for the technologist is a significant service to the laboratory and is another example of how the computer can be used as a tool. Pribor and associates (1968b) perform the calculation of electro-phoresis results after direct acquisition of the data from the scanning instrument.

Other nonautomated applications include the filing of laboratory data and retrieval for the preparation of histograms and standard-deviation calculations (Hicks et al., 1966). In general, only programs such as calculations which are required throughout the day have been implemented to run simultaneously with on-line monitoring. Programs to perform statistical calculations, for example, are used only when the computer is not running on-line. The nonautomated off-line application was the first use of a LINC in a clinical laboratory and has been described elsewhere by Hicks et al. (1966).

VI. EXPERIENCE WITH ON-LINE DATA

A. Results

The real test of the system is, of course, whether or not it can consistently and reliably obtain the correct answers when compared with the procedures used by laboratory personnel. The on-line system has been designed to emulate the technologist. In the same manner that the technologist runs standards on her instrument and constructs a calibration curve, the LINC also uses the voltage readings to construct a point-by-point curve. Linear interpolation between standard points on the curve has given consistently good agreement between manual and on-line results (Evenson et al., 1967). The more standards that are included in the curve, the better is the accuracy of the results. The entire working range of the curve does not need to be linear. Data from 12 different instruments operated on-line has shown that the accuracy of the result from the LINC is improved in comparison to the manual reading method.

Two sources of error—baseline drift and changes in sensitivity—must be corrected frequently in most automated instruments. Both are corrected by frequent complete restandardization of the instrument. When the computer is doing the calculations, there is an incentive to calibrate more frequently, thereby improving performance. With the new techniques of Habig et al. (1968) to increase the speed of automated equipment many times there is more than adequate time to standardize the instrument frequently.

Another test of the system is whether or not it is used by the laboratory. If laboratory personnel will not use the computer because "it is easier to do it myself" or if a complete dependence upon computer experts is developed, then by definition of the philosophy stated in this report the system is a failure. Experience in several laboratories has proved that the system can be operated by laboratory personnel. The technologists are able to implement new tests and make changes in existing tests without becoming programmers or computer experts. In a laboratory with large volumes of automated tests, the enthusiasm of the laboratory personnel for the computer is proportional to the amount of work the computer saves.

B. Real-Time Quality Control

A major advantage of using a computer on-line in real time with instruments is to provide an interactive quality control program in the laboratory. By continuously monitoring equipment in the laboratory, the computer can sense malfunctions and errors in results within minutes after they occur. If the technologist is notified immediately as problems occur, action can be taken in the laboratory before a significant amount of data is lost or time is wasted. Thus it actually happens that a technologist may shut down an instrument on the basis of computer diagnostic printouts when the computer has printed many diagnostics indicating that errors are occurring in the on-line system. The computer can often sense subtle changes in instruments before they are noticed by casual observation of the chart recordings.

The basis for the detection and communication of instrument changes is the reading diagnostic pattern printed on reports. The series of 1's and 0's gives the reading pattern for the points taken on each peak from the instrument. The principle for the generation of this pattern is shown in Fig. 1-12. In the operation of the laboratory monitor program, after the

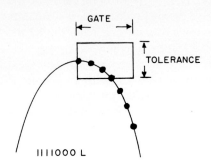

Fig. 1-12. Principle of reading diagnostics.

computer has been synchronized on the first peak, the computer takes readings at regular time intervals. For example, on an instrument that runs at 60 samples/hr the computer takes a series of readings every 60 sec after the switch is turned on at the first peak. The number of observations or readings taken at 1-sec intervals on each peak establishes the length of a time gate, i.e., the time during which the peak is observed as shown in Fig. 1-12. The tolerance establishes the allowable scatter of the readings taken on each peak during the time gate. The dimensions of the gate and tolerance can be envisioned as a rectangle that can be placed directly over a peak as shown in Fig. 1-12. The leading edge of the rectangle is placed on the peak at the point at which the first reading is taken. From that point on, readings are simply taken at 1-sec intervals. All readings that fall inside the box are said to be "in tolerance." Those readings that fall outside the box are obviously outside the tolerance limits. Thus the diagnostic generated by the situation illustrated in Fig. 1-12 would be four 1's followed by three 0's. The L following the reading diagnostic means that out-of-tolerance readings are low.

A series of readings taken on a peak can be treated in two different ways by the laboratory monitor program. First, LABMON can simply select the highest reading as the peak and use that reading as the voltage value. In this case the highest reading is always taken as an acceptable value and the tolerance of all of the other readings are compared with the highest reading. Any reading that is out of tolerance will always be low, since the highest reading is always taken as an acceptable reading. A second method of treating these readings is to average all of the observations and determine the scatter of the readings with respect to the average. In this case some readings can be out of tolerance with high values and some with low values. A laboratory study of these two methods of treating the observations has been made by Evenson *et al.* (1968) and has revealed the following conclusions.

In general, it has been concluded that both the "peak top value" method and "peak mean" method give essentially the same results. As the operation of the instrumentation becomes marginal, however, the meanings of the diagnostics are somewhat different. The peak-mean technique is, for example, more sensitive to the rate of change of pen position on the chart recorder and therefore provides more information as to the range of readings taken during the reading interval. With the peak-mean technique, the tolerance can be made smaller, and the technique is more sensitive to "noise" in the on-line system. Further, by averaging the observations on each peak, the readings are less sensitive to changes in concentration from one sample peak to another. Always taking the highest value can on occasion lead to an erroneous observation. This is especially true of techniques for improving the speed, accuracy, and precision of Auto-Analyzer systems.

The reading diagnostics provide the technologist with a good "picture" of just how the computer "sees" each peak or set of readings on an instrument. Figure 1-13 is a printout demonstrating a problem with an instrument. In this case numerous reading diagnostics were generated for the urea nitrogen chemical analyzer. Further inspection reveals that the diagnostics alternate from those with 1's on the leading edge to 1's on the trailing edge of the peak. This characteristic alternating of the diagnostic readings means that the peak on the instrument is "wobbling" or moving erratically within the timed reading gate. Even with this number of diagnostics on the computer printout, direct observation of the chart does not easily reveal this problem. In the particular instance shown in Fig. 1-13, a leaking dialyzer membrane was discovered when the instrument was shut down and checked. Other characteristics such as sample interaction, leaks in tubing, and clots in the sample aspirator also produce diagnostics. Any noise from reagents in the chemical analyzer or from the electronics in the interface will also cause diagnostics to be generated. In general, a reading diagnostic with alternating 1's and 0's or "holes" in it indicates a source of noise. Watching the analog channels on-line with the display channel program usually reveals the source of the noise.

These are a few examples of the subtle observations that can be made by the on-line computer system. With the further refinement of diagnostic

```
UREA NIT          03--05--68
 1   STD      5               138
 2   STD     10  0000011 L    175
 3   STD     30               281
 4   STD     50  1100000 L    364
 5   STD     70  1110000 L    418
 6   STD    100               473
 7   STD    150  1111000 L    505
 8   POOL    20  1100000 L    231  OUTSIDE LIMITS [   21 -   25 ]
 9           14  0000001 L    200
10           69  1100000 L    417
11           34  1100000 L    299
12           11  0000001 L    184
13           51  0000001 L    369
14       EXCEEDS RANGE
15           62  1100000 L    399
16           14  1110000 L    199
17           10  0000011 L    177
18           37  1100000 L    311
19           13
20           19  1111100 L    225
21           21  1111000 L    237
22           24  1110000 L    253
23   VERS    12  OUTSIDE LIMITS [   13 -   14 ]
```

FIG. 1-13. Report showing use of diagnostics.

techniques and methods of making observations on signals produced by instruments, it should be possible to further increase the value of the quality-control system. A very important advantage of making these observations in real time in the laboratory is that it is possible to take corrective action at the time that the problem occurs. The collection of readings from an instrument followed by off-line processing of the results several hours later can never match the real-time quality-control capability of an on-line computer.

C. Capacity of System

The first on-line system was designed on the classic LINC computer having only 2000 words of memory. With this small memory, only one other program can usually be run in the computer with the LABMON system as shown in Fig. 1-3. Consequently, when the computer is printing on-line reports it cannot be simultaneously processing more data from the on-line system. Also, when a technologist is using the computer to perform a manual calculation, the computer cannot continue to print on-line reports until the technologist has completed the job. With this configuration, when about a dozen analog channels are operating simultaneously on the computer, including the SMA-12 AutoAnalyzer (giving a total of 800 to 900 test results per hour to be processed), the on-line processing and printout programs begin to lag behind the laboratory because of the slow speeds of the printing and magnetic-tape operations. That is, with each 10-min period that passes in the laboratory, the on-line reports fall a little farther behind and cannot keep up with all of the results being generated on a real-time basis. In the worst case, the reports may continue to be printed for one to two hours without stopping until the system catches up. The computer, of course, never lags behind in reading the instruments. While delayed printout of results does not in itself present a serious limitation, it does prevent the computer from being used for any other operations except on-line monitoring while the reports are behind.

There are several solutions to this particular problem. It is estimated that about 0.2% of the computer's time is required to keep a teletype printing at top speed. However, with only 2000 words of memory there is not enough space to do anything more while the teletype prints at its slow pace. With even a modest increase in memory size it is possible for the LINC to keep the teletype printing reports while simultaneously performing other operations. Another major problem with earlier systems is the magnetic-tape speed. During the processing of on-line data, almost all of the time required can be accounted for by the searching for data on the magnetic tape. This observation is made very easily. On the console of each LINC there is a light that turns on when the central processing unit is functioning. During the time the computer is waiting for data to be transferred from the magnetic tape, this light is off. During the entire procedure of processing on-line data, the LINC "run light" is seldom on because the computer never runs long enough for the filaments in the light bulb to heat up. Buffering the magnetic tapes from the central processing unit can be of some small help. By buffering the tapes, the central processing unit can continue to process data while the magnetic tapes are being searched for more information. However, the processer is so fast that the calculations are performed almost instantaneously and the processer still spends essentially all of its time waiting for more information from the tapes. The use of high-speed disc files on the LINC for most of the on-line data offers a convenient solution to this particular problem. The disc on the LINC has an average access time of 20 msec as opposed to several seconds for tape. The high speed of the disc files, combined with the high speed of the central processing unit, gives rapid processing of data. The use of high-speed disc files and the complete buffering of the teletype terminals so that the central processer can be used for other programs while reports are being printed permits the computer to keep up with all other on-line operations and perform many new tasks. The actual memory speed of the central processing unit has yet to be a limiting factor in any of the systems developed to date.

VII. CRITIQUE OF SYSTEM

Several lessons have been learned through experience with on-line operation in the clinical laboratory. In the first place the connection of automated instruments directly to the computer has required extensive planning of the communication between the machine and personnel. Not only is it necessary to communicate with the personnel operating the instrument, but the computer must monitor the equipment the person is using and check all requests and responses against the operation of the equipment. At first it seems that one advantage in dealing with automated systems is that it reduces the amount of human intervention. Unfortunately this is the case only when instruments work perfectly, and this seems to be the exception rather than the rule. Furthermore, the operation of the clinical laboratory is constantly complicated by new and changing procedures and emergency requests that can be intercepted and reacted to only by human beings.

Another lesson learned with the on-line system is to give proper attention to even the smallest detail. For example, the logistics of operating the switch on the instrument can cause more work for a technologist than ever existed before if it is not properly implemented. An apparently simple procedure such as automatically turning the switch off after 40 cups makes the difference between efficient production of results from an instrument all day long and running back and forth to the computer to constantly communicate what is going to be done next. There is a seemingly endless number of these details that must be dealt with, including even the language used in the conversational mode with the technologist. The word *schedule*, for example, means many different things to many people.

The on-line system has offered an excellent opportunity for laboratory personnel to become acquainted with the computer because the system is operated on a real-time basis by all laboratory personnel without the assistance of professional programmers. The requirement that a computer perform all day every day without a failure is a very demanding one. However, an acceptable degree of reliability can be achieved once the programs have been tested for all of the combinations of events that occur in an operating clinical laboratory.

The addition of more memory and a high-speed disc file to the LINC system permits a great expansion within the laboratory. With the new high-speed system, even with 24 analog channels in operation, there is still sufficient computer time left to implement administrative procedures to identify on-line data and maintain patient files. As stated previously, the next phase of development is to identify all data in the data-acquisition system to create a patient file structure. By implementing multiple teletype terminals to communicate with the computer simultaneously, the workload of entering administrative information can be distributed throughout the laboratory. This organization of the system permits the next step to be taken.

The major limitations of the system have been concerned with the ability of a small computer to keep up with a dozen or more analog channels producing data while having enough time left to implement administrative procedures. The expansion of memory and the addition of disc files are solutions to these problems. The major advantages of the system continue to be economy, maintenance of control of the computer, and convenience of use in the laboratory. For laboratories with a sufficient amount of automation, a computer can pay for itself easily just by performing all of the calculations from the automated equipment. This has been demonstrated at several installations.

VIII. EVALUATION

On-line data acquisition is a very large step toward a total information system. Even in the collection and organization of information on-line, it is necessary to develop organized file systems. When it comes time to structure the "patient file," most of the information has already been collected and the step of identifying the information permits the patient file to be generated. For information that is not generated automatically, the use of the conversational mode permits the data to be entered directly into the computer.

The computer has been of significant value in acquainting laboratory personnel with the new technology. The computer has been accepted as a common tool in the laboratory used by all personnel. The reliability of the computer exceeds that of any other instrument in the laboratory. Failures certainly do occur, but at a much lower frequency than in any other instrument. The currently acceptable level of failures seems to be about one per month (of a nature that causes a significant loss of information). Major problems with the computer system continue to be the mechanical devices such as magnetic tapes and teletypes. The experience gained in learning how to fill out a trouble report and how to report that a computer failure has occurred has been invaluable. Through the development of well-defined trouble-reporting procedures, it is possible for laboratory personnel to preserve adequate information to solve most problems after they have occurred.

The capacity of the on-line system appears to be quite adequate for 24 analog channels. This should permit installations in hospitals of up to 1000 beds. Implementation of the computer system through the use of user-control programs has proved to be a simple procedure that can be mastered by laboratories without the assistance of programmers. Upon implementing the on-line system, laboratory personnel become intimately acquainted with the computer in the same way that they become acquainted with chemical analyzers and can thereby maintain the computer system in good operation from day to day. Just as in the case of all laboratory equipment, the better laboratory personnel understand the system, the better it operates.

One of the most important advantages of the LINC continues to be its ability to grow. With the addition of disc files and expanded memories, the LINC still continues to be a LINC. All programs written still continue to run on the same equipment, and additional peripherals can be added as necessary. As the laboratory grows, so can the LINC grow into a larger and larger computer system.

In general, by molding the computer to fit into the laboratory's problems, no compromises have been made with laboratory procedures simply for the sake of using the computer. The chemical procedures are determined by chemists, not programmers. When a technologist wants to change a pool control sample or add more standards to the chemical procedure, it is not necessary to consult a programmer. This degree of flexibility has made the system highly acceptable to the laboratory.

There is a close analogy between the development of computer methodology for the clinical laboratory and the development of chemical methodology for chemical procedures. In the on-line system, where there is a virtual marriage between the computer and the procedure in the laboratory, the computer can greatly limit the flexibility of a test procedure if implemented improperly. Through the mechanism of continued critical evaluation by the field of clinical laboratory medicine in the same sense that methodology has been reviewed, it should be possible to develop a laboratory information system in the future that is stable and acceptable to the laboratory community and medicine in general.

Acknowledgment

This work was supported in part by National Institutes of Health Research Grant No. FR 00249 from the Division of Research Facilities and Resources. The cooperation of the Wisconsin Clinical Laboratory staff and the technical and programming assistance of Mr. Robert Carr are gratefully acknowledged. The authors also appreciate the continued encouragement and support of Mr. Edward Connors and Mr. John Russell of the hospital administration.

References

Blaedel, W. J., and G. P. Hicks. (1964). *Advan. Anal. Chem. Instr.* **3**, 105.
Blaedel, W. J., and R. H. Laessig. (1966). *Advan. Anal. Chem. Instr.* **5**, 69.
Blaivas, M. A. (1965). *Mod. Hosp.* **104**, 116.

Brecher, G., and W. H. Wattenburg. (1968). *Am. J. Clin. Pathol.* **49**, 248.
Clark, M. A. (1967). LAP6 Handbook, Technical Report No. 2, Computer Research Laboratory, Washington University, St. Louis, Missouri.
Clark, W. A., and C. E. Molnar. (1964). *Ann. N.Y. Acad. Sci.* **115**, 653.
Cotlove, E. (1965). *Clin. Chem.* **11**, 816.
Evenson, M. A., G. P. Hicks, J. A. Keenan, and F. C. Larson. (1967). *Automat. Anal. Chem.* **1**, 137.
Evenson, M. A., G. P. Hicks, and R. E. Thiers. (1968). *Clin. Chem.* In press.
Flynn, F. V., K. A. Piper, and P. K. Roberts. (1966). *J. Clin. Pathol.* **19**, 633.
Giesler, P. H., and G. Z. Williams. (1963). *Hosp. Top.*, Feb., 73.
Habig, R. L., B. M. Schlein, L. Walters, and R. E. Thiers. (1968). *Clin. Chem.* In press.
Hicks, G. P., W. V. Slack, M. M. Gieschen, and F. C. Larson. (1966). *J. Am. Med. Assoc.* **196**, 973.
Lamson, B. G. (1965). *In* "Computers in Biomedical Research" (R. W. Stacy, and B. Waxman, eds.), Vol. I, pp. 353–376. Academic Press, New York.
Lindberg, D. A. B. (1965). *Missouri Med.*, April, 296.
Peacock, A. C., S. L. Bunting, D. Brewer, and G. Z. Williams. (1965). *Clin. Chem.* **11**, 595.
Pribor, H. C., W. R. Kirkham, and R. S. Hoyt. (1968a). *Mod. Hosp.* **110**, 104.
Pribor, H. C., W. R. Kirkham, and G. E. Fellows. (1968b). *Am. J. Clin. Pathol.* **50**, 67.
Rappaport, A. E., W. D. Gennaro, and W. J. Constandse. (1968). *Mod. Hosp.* **110**, 94.
Schoop, R. A., B. Klionsky, and J. S. Amento. (1968). *Clin. Chem.* **14**, 197.
Slack, W. V., and L. J. Van Cura. (1968a). *Postgrad. Med.* **43**, 68.
Slack, W. V., and L. J. Van Cura. (1968b). *Comp. Biomed. Res.* **1**, 527.
Slack, W. V., G. P. Hicks, C. E. Reed, and L. J. Van Cura. (1966). *New Engl. J. Med.* **274**, 194.
Slack, W. V., B. M. Peckham, L. J. Van Cura, and W. F. Carr. (1967). *J. Am. Med. Assoc.* **200**, 224.
Straumfjord, J. V., M. N. Spraberry, H. A. Biggs, and T. Noto. (1967). *Am. J. Clin. Pathol.* **47**, 661.
Thiers, R. E., R. R. Cole, and W. J. Kirsch. (1967). *Clin. Chem.* **13**, 451.
Williams, G. Z. (1964). *Military Med.* **129**, 502.
Williams, G. Z. (1967). *Bull. Coll. Am. Pathol.* **21**, 383.

CHAPTER 2

Automated Multiaccess System for Clinical Work[1,2]

WILLIAM J. MUELLER

STATE UNIVERSITY OF NEW YORK, UPSTATE MEDICAL CENTER, SYRACUSE, NEW YORK

I. Problems Associated with the Automation of Clinical Areas ..	55
II. System Implementation..............................	59
A. Hardware	59
B. System Operation and Software	69
III. System Performance	74
IV. Current Clinical Experience	79
V. Critique	79
VI. Evaluation	82
Reference	85

I. PROBLEMS ASSOCIATED WITH THE AUTOMATION OF CLINICAL AREAS

Our department of Clinical Pathology, like similar departments in other hospitals, has an increasing work load and a shortage of competent technicians. At their request, members of the Department of Bioelectronics and Computer Sciences investigated their operation for possible improvement through the use of our computer facility. The following paragraphs describe the procedures that were in use and some of the problems associated with automating them.

---

[1] The computer system described in this chapter was developed in collaboration with the members of the Department of Bioelectronics and Computer Sciences and the physicians, nurses, and administrative staff of the State University Hospital.
[2] Supported by a NIH Special Research Resources Grant FR 00353 for the central and satellite SEL 810A computers and by NIH National Institutes of General Medical Sciences Grant GM 11413 for the development of BASP and the computer display and interface devices.

Starting with the doctor's written order for a clinical test, there were six transcriptions of that order before a report of the test results was returned to the physician. The nurse at the ward interpreted the doctor's order and filled in a request slip, which was sent to Clinical Pathology. Here the technicians generated the work sheets, carried out the required tests, and filled out daily logs.

We found that a major requirement of a data-processing system for the collection and distribution of clinical laboratory results is a device to automatically enter the data from AutoAnalyzers to the computer. As data generated by these analyzers represents approximately 80% of the test results of a clinical laboratory, such a device would solve a large portion of the data-entry problems involved in automating the clinical laboratory.

Figure 2-1 is a section of a typical AutoAnalyzer record. This record shows a series of curves, the peak of each curve being the analyzer result. Technicians were required to list the peak values of the curves, calculate

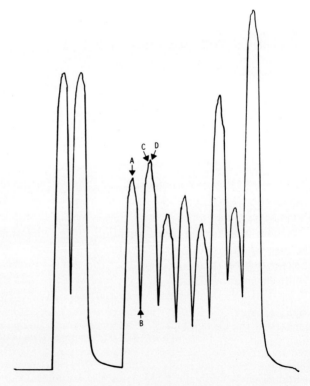

FIG. 2-1. A typical AutoAnalyzer record.

2. Automated Multiaccess System for Clinical Work 57

the test result by comparing the peak value to standardization peaks, normalize to the proper dimensions (milliequivalents per liter, for example) and record the test result in the patient record. An instrument to automate this procedure would be required to select and digitize peak-value points such as point A, recognize point B as the start of a new sample, and disregard point C as a false peak but accept point D as a correct peak.

Tests made using general devices, such as pH meters and spectrophotometers, were also recorded by the technician doing the test and later transcribed to a cumulative report sheet for the patients' records. In general, about 30% of technicians' time was used for nonproductive paper work with a high probability of transcription error.

Over a hundred phone calls were made daily by physicians to the Clinical Pathology Department to check on clinical tests. Each phone call required a secretary to locate the technician who had the sample for that test and to determine the status of the test. This information, which took two to three minutes to obtain, was then relayed to the physician.

At 5:00 p.m. each day the test results on each patient were copied from the daily log, duplicated, and sent to the ward for insertion in the patient's hospital record.

With 250 in-patients requiring 1200 tests, and with an expansion program to provide a 500-bed hospital past the planning stage, it was apparent that some form of automation in Clinical Pathology and improvement in communications between that department and the physician was required.

An attempt was made to improve the situation by having a keypunch operator generate punched cards for each test requested and then, using an available computer (CDC 160A), produce the work sheets, daily logs, and cumulative reports. However, the card punching took three to four hours, requiring the keypunch operator to start work at 3:00 a.m. The CDC 160A computer, being partly vacuum-tube operated, required a warm-up period for stable operation so the computer personnel had to begin work at 7:00 a.m. Neither condition was well-tolerated, and as the reduction in paper work did not justify the effort required, the program was stopped and an investigation was begun to establish a procedure that would eliminate the tedious card-punching operation.

While we were having meetings with the personnel of the Clinical Pathology Department, we began receiving requests from other service departments that wanted to display current information and provide long-term storage for later retrieval and analysis. In effect we were requested to develop the nucleus of a hospital information system.

Because each department operated independently and had its own reporting procedures, we began an intensive study of the hospital from the admissions office to the discharge area. We found that each clinical service section or department (Clinical Pathology, EEG, ECG, Pathology, Pharmacy, and Radiology) had a different test-request procedure. The one common denominator in all test requests was a charge card that notified the business office that a test request had been made. One of the most difficult problems was to develop a computer-acceptable test request procedure that would be accepted by the hospital personnel involved and by the business office for billing purposes.

Physicians at this time also began inquiring about the feasibility of on-line processing of patient analog signals, especially during surgical procedures. Several of their requested analyses had been done by others and are similar to those described in other chapters of this book. However, some surgical procedures were described to us in which on-line averaging and correlating of data from a multielectrode array in a patient's brain would reduce the risk of destroying critical areas. Also, physicians doing ECG analysis proposed ways in which complex on-line monitoring and analysis could predict an impending malfunction of the heart. We found that some programs for on-line analysis by a standard computer took a considerable amount of time to write, and if the analog data sampling rate was faster than 1000 samples per second it was very difficult to operate any medium-sized computer system in an interrupt mode for more than three patient-data channels while servicing Clinical Pathology and other patient-analysis areas.

The problems, then, were to:

(a) Develop a fast method of entering clinical-test requests into the computer and have that procedure acceptable to the clinical and administrative personnel.

(b) Modify AutoAnalyzers to provide direct data entry into a computer.

(c) Generate all necessary work sheets, logs, and reports and store test results for later retrieval and statistical analysis.

(d) Accept data from all clinical service departments for long-term storage on magnetic tape.

(e) Provide an immediate display, on demand by the physician, of all test requests and results, and other pertinent information such as drugs requested and x-ray reports.

(f) Have a paper trail to enable the system to function in the event of an equipment failure and to provide the business office with a list of charges.

(g) Design the system to be on-line, reliable, easily expanded, and optimized on a hardware–software basis for best cost-to-performance ratio.

(h) Develop a preprocessor to provide high-speed multichannel analyses

2. Automated Multiaccess System for Clinical Work 59

of such complexity as to preclude doing them on the central computer operating in an on-line, real-time environment.

(i) Provide batch processing of patient data and data from research animals, for physicians and investigators, on an interrupt basis, while handling data that are being collected for long-term storage or for off-line analyses.

II. SYSTEM IMPLEMENTATION

A. Hardware

It seemed obvious that one of the third-generation process-control type computers, operating in a foreground–background mode with multiple interrupt capability, would be ideal for the problems we had. This type of operation lets the computer operate as a batch processor in the background until interrupted by a foreground program to service an external on-line device.

We contacted six manufacturers of such equipment and did an analysis of their proposed systems based on the speed of the central processor and peripheral units, the method used for keeping track of the foreground–background operation, the number of interrupts and manner in which interrupts were handled, and the cost of a system matching as closely as possible a set of hardware specifications that represented the average of the systems proposed. As was expected, the most sophisticated systems had the highest cost and required the most memory (generally 8–16 K words) for nonproductive supervisory work and 16 K for normal program operation. To add an additional channel and therefore another interrupt, the system had to be taken out of operation while a modified supervisory system was read in. Other systems required the addition of a conditioner-interrupt card to set up another channel. In all cases, if the number of input devices was of any size (say, fifty units), the complexity of programming became extremely time consuming or the input-conditioning cards cost more than the central processor.

Other systems that used cards and teletype machines or multiple small processors were also investigated. In each case we could not approach the operating system we believed necessary, and the cost was higher than could be justified by a reduction in personnel requirements.

Further, trial clinical programs were run on paper and the time required by the several computer systems was calculated. In some instances the correction of test results for drift of an AutoAnalyzer and their storage on a disk required from 12 to 60 seconds, during which time the computer was unavailable for other work. An interrupt could have been made by a higher-priority signal, but if the signal was another AutoAnalyzer the problem still existed. Even though the data rate was slow on each channel, chance alone (because all input devices are asynchronous with the computer) could cause an interrupt sequence that the system could not service.

Our calculations showed that any computer we investigated, when operating in a foreground-background mode, would have less than 30% of its time available for batch processing while doing the necessary clinical work. The continuous shuffling of data between the CPU and the disk memory, caused by having a large number of interrupts, is very time consuming and of course makes the cost per test high. To reduce the cost per test, it was necessary to reduce the amount of computer time dedicated to the calculation and storage of clinical tests and leave more time for batch processing or some other operation such as patient monitoring.

As we were unable to find a standard computer system that operated in a foreground–background mode under a time-shared monitor system that could do what we believed desirable at a reasonable cost, we designed a somewhat unconventional system. At this point in our investigation we were developing a small Biomedical Analog Signal Processor (BASP), which was to be used for time-shared analyses of analog signals from

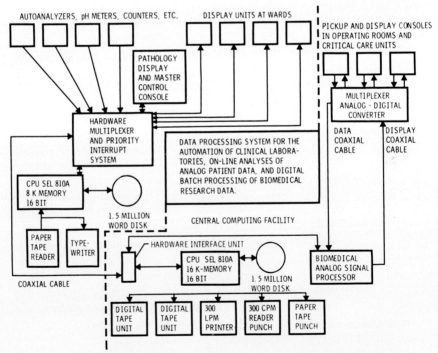

FIG. 2-2. Diagram of satellite and central computer system.

experimental animals. The time-shared system was hardware-controlled, required no supervisor program, and was fast and low in cost. Using some of the time-sharing concepts of BASP, the computer system shown in Fig. 2-2 was developed.

A satellite computer consisting of a Systems Engineering Laboratories (SEL) 810A central processing unit (CPU) with an 8 K, 1.6-μsec, 18-bit memory, an ASR33 Teletype, and a 1.5-million-word disk memory, is in the hospital and dedicated to the on-line collection and dissemination of clinical test results. A central computing facility is in our Medical School. This facility consists of an SEL 810A CPU with a 16 K, 1.6-μsec, 18-bit memory and a teletype. Peripheral units are a disk memory identical to that in the satellite unit, two 45-kHz 800-bit-per-inch (bpi) digital tape units, a 200-card-per-minute (cpm) card reader, a 300-line-per-minute (lpm) printer and an 8×11-in. plotter. Each CPU has parity check, memory protection, and power-failure protection. Both systems have a digital input/output unit to permit data to be entered from, or transmitted to, external devices on an interrupt command.

A multiplexer and a delay-line memory (Fig. 2-3) are at the input to the digital I/O unit of the satellite computer. The multiplexer is continuously rotating, one step each millisecond, through 64 input channels that consist of three 16-bit computer words. If the external device (an AutoAnalyzer, for example) has turned on a flag bit, the multiplexer, upon switching to that channel, dumps the three words into the delay-line memory, using a parallel-to-serial converter. The flag bit of that analyzer is reset to zero, and the multiplexer, after a 25-msec delay during which data are entered into the memory, indexes to the next device. If the flag bit is "on," the three words of data from the second channel are dumped into the delay-line memory behind the first three words. On the other hand, if the second flag bit is a zero, the multiplexer skips to the next device in a millisecond.

It should be noted that this is a hardware system that operates independently of the computer. The successive channels of data are stacked behind the first set of data in the delay-line memory until a time when the computer is not busy. The computer accepts the first three words from the delay-line memory and proceeds to process them. This procedure removes the first three words from the delay-line memory, and the second three words move into first place. Thus the memory works on a first-in, first-out basis and provides a buffer between the asynchronous data sources and the computer. Because of the slow data-collection rate the delay-line memory, with a capacity of 512 words, provides an effective time delay of about two minutes, during which the computer can do any necessary operation without being interrupted.

This combination of a hardware multiplexer and a 512-word delay-line

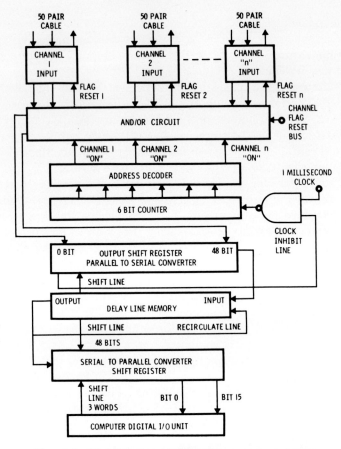

Fig. 2-3. Multiplexer and delay-line memory system.

memory effectively replaces a supervisor program and the 8–16 K core memory required for interrupt operation. The system is hardware-protected and can be easily expanded by letting the multiplexer rotate through more positions. A limit will be reached when the data-input rate exceeds the capabilities of the computer with the buffer—that is, when it takes longer to process data than it takes to fill the buffer. At this time another delay line memory can be inserted behind the first and another hundred or more channels can be added.

A second multiplexer (Fig. 2-4) is used to sequentially scan the keyboard–display consoles in the various service departments and the display consoles on the wards. The keyboard–display consoles can modify the data stored in the computer, whereas the display consoles on the wards can only receive requested information.

2. Automated Multiaccess System for Clinical Work

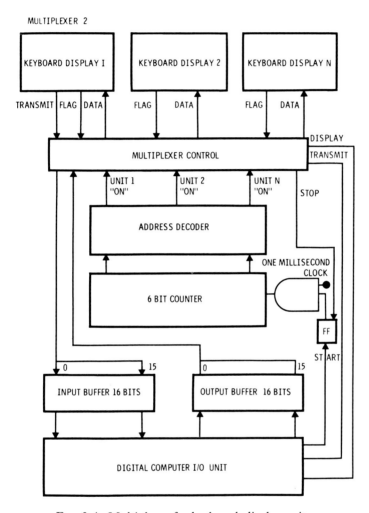

Fig. 2-4. Multiplexer for keyboard–display units.

The second multiplexer has a control system that interfaces the computer digital I/O unit, through the hardware priority-interrupt system, to the keyboard–display or display-only units. When a flag is raised on a display unit, the multiplexer control checks the transmit or display bit to determine which operation is desired and stops the clock.

If a patient's hospital number has been dialed on a display unit and the "display" button pushed, the hospital number is dumped into the digital I/O unit along with the display bit. This causes the computer to search the disk for the requested patient's page and to transmit the page, two characters at a time, through the output buffer to the display-unit delay-line memory. The transmission of up to 1600 characters takes 16 msec, which is one rotation of the memory. The computer sends a data complete signal and the clock of the multiplexer is started.

If a "transmit" button on a keyboard–display console has been pushed, the clock is stopped and the patient's hospital number and the transmit bit are dumped into the digital I/O unit. This causes the computer to send a "ready" signal, which, when the data have rotated to the starting position in the delay-line memory, initiates the transmission of a page from the console memory through the buffer input of the digital I/O unit to the disk memory of the satellite computer. This data page replaces the data page originally assigned to the patient's number. A "data received" signal is sent by the computer, and the clock is started so that the next display device is checked for a raised flag.

The display multiplexer is arranged to have priority over the first multiplexer-memory unit, so that display devices are serviced almost instantly while the much slower analyzers wait their turn.

Our initial trials are being made with one keyboard–display console in Clinical Pathology and two display consoles that are being rotated through the wards, The display consoles are experimental CRT units of our design and construction and display 800 raster-type characters, which are obtained from a monoscope character generator.

Provisions are made for refreshing the display, using an inexpensive delay line and some logic modules that decode special signals on the delay line to determine the beginning and end of data. A system of counters and shift registers permits data to be entered or transmitted from the delay-line memory at a rate of two characters every 16 msec or slower (e.g., for keyboard entry) or at a rate of 1600 characters in 16 msec for display and entry to the computer.

Eight rotary switches on the front of the display unit are used to dial a six-digit hospital number to identify the patient and a two-digit number identifying the particular tests that are to be displayed. A "display" pushbutton raises a flag bit, which interrupts multiplexer number 2 and through it requests service from the computer.

The pages assigned for storing tests for different departments are not all the same length, and indicator arrows on the side of the display CRT show the operator the maximum number of lines that are available for data storage. The arrows change position with the setting of the two code switches that call a particular test page. Any data entered below the arrows is truncated when sent to the computer.

In the keyboard–display unit, data can be entered, modified, and transmitted to the computer in a truncated ASCII code. A cursor is used to indicate which character is to be inserted, modified, or erased. Data can be entered directly into the delay-line memory from the incremental card reader for display, modification, and transmission to a patient's page on the disk of the satellite computer.

There are hardware parallel-to-serial and serial-to-parallel shift registers and line-driver systems at the satellite and central computers to provide an interface, via coaxial cable, between the two systems.

Initial patient data are entered by a hand-fed incremental card reader at the keyboard–display console. The card is moved two columns, and therefore two characters, for each read instruction. This permits the entry of the card data to the display unit two characters at a time. At approximately 16 msec for two characters, it takes $\frac{1}{2}$ sec to read a card. Our test-request procedure uses a punched card with a carbon copy of the doctor's order glued over the holes in the card. Because of this, we designed a card reader that photoelectrically reads the absence of light where a hole is punched. The face of the card is in an enclosed black chamber, while the bottom of the card is illuminated and scanned by twelve photocells. Standard Hollerith code is converted to ASCII before computer entry.

The preprocessor BASP is connected as a peripheral unit to the central computer to provide high-speed processing of analog signals while slow-speed processing is done by the central computer. Analog data are digitized and sent to BASP through a remotely controlled multiplexer and analog-to-digital converter unit placed at the point of origin of analog signals. We prefer digital transmission of data and use spare coaxial cables that were installed when we put our extensive closed-circuit color television system in operation two years ago.

A 1-MHz clock is delayed at the central facility and then aligned with the incoming digitized data, which have been delayed by the transmission time from the origination point through the coaxial cable, which in our case may be a mile long.

Various methods for displaying computed results are being investigated, but at this time we prefer a storage oscilloscope for graphical display and IEE numerical displays of the CRT type for digital readout. The CRT displays can be read at a distance, have all numerals in the same plane, and have high brightness to permit viewing in well-illuminated areas.

The philosophy behind BASP has been published elsewhere (Mueller et al., 1968) and need not be repeated here. It should be noted, however, that the preprocessor is not programmable in the usual sense and cannot do a large number of simple operations such as one would use, for example, to do a standard ECG analysis. Rather, the device is designed to be easily programmed from a flow chart and to do operations involving at most two or three add–subtract, multiply–divide, square–square root, integration, or core read–write cycles per calculation at a data point on an analog signal. The same calculations are done on all data points that follow—that is, the program does not change with time. This is the manner in which on-line averagers, correlators, variance analyzers, and some model-building systems work. The high speed (1-million data points per second) permits multichannel operation, with a different program on each channel if desired. But only a limited number of data points (1024) can be held for each of eight time-shared channels. Initial trials have shown this to be adequate. More data points can be used, but the cost of high-speed core memories places a limit on the number of channels and data points that can be in simultaneous operation. One can, of course, use more channels and fewer data points per channel.

To permit the on-line entry of data from AutoAnalyzers we have designed and constructed an instrument capable of selecting the desired peaks from the AutoAnalyzer curves of Fig. 1 and preparing them for computer entry. Figure 2-5 is a block diagram of the device. The input for the unit is obtained from a retransmitting slide-wire assembly mechanically attached to the analyzer recorder slide-wire assembly. This system gives an output

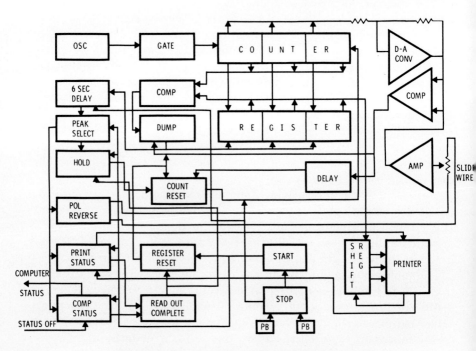

FIG. 2-5. AutoAnalyzer–computer interface unit.

2. Automated Multiaccess System for Clinical Work   67

of 0 to 5 V corresponding to the analyzer recorder pen position. An operational amplifier wired for unity gain uses this voltage as input and has its output wired to one input of an analog voltage comparator. The other comparator input is a staircase voltage derived from a digital-to-analog (D–A) converter wired to a three-decade binary-coded-decimal (BCD) counter. An oscillator drives the counter and the D–A converter to give an output of 5 mV per count for a resolution of 0.1%. When the output of the D–A converter equals the analog input voltage from the analyzer, the comparator generates a signal. At this time the BCD value in the counter is proportional to the analyzer pen position. The signal generated by the comparator is used to reset the counter, and the sequence just described starts again.

A three-decade register is wired through logic gates to the BCD counter. The output of this register and the counter are wired to a digital comparison circuit. Any time the counter and register are equal, signals are generated that will, through the logic gates, cause the register to follow the counter. When the D–A converter reaches the analog input voltage, the signal generated by the analog comparator immediately disconnects the register from the counter and then resets the counter. This action stores the counter value in the register and resets the counter to allow it to look for a new value. If the new value is larger than the value stored in the register, the counter (on its way up) must reach comparison with the register. When this happens, the register will follow the counter to the new value. If the new value is smaller than the register value, the counter will not connect to the register and the register value will not be affected. Eventually, the highest point reached by the counter, equivalent to point A of Fig. 2-1, will be stored in the register.

When the digital value of the highest point on the curve is stored in the register, digital comparison signals can no longer be generated. The lack of these signals for a 6-sec period is interpreted by the instrument as meaning that the value stored in the register is the real peak value and not a false peak such as point C of Fig. 2-1. At this time a hold signal is generated that (1) holds the counter in reset, (2) raises a flag bit indicating that data are ready for transfer, (3) reverses the polarity of the slide-wire reference voltage, causing point B of Fig. 2-1 to appear to the instrument as a maximum point instead of a minimum point, and (4) generates a "print" command signal.

The "print" command signal is used as input to a shift register that, in conjunction with a printout system, shifts the contents of the register digit by digit to the printer. When the printer completes the register printout, it sends a "print complete" signal back to the instrument. This signal, along with a signal sent from the multiplexer memory unit when it has completed input of the register data, resets the register to zero and releases the counter to look for a new peak.

Because (as mentioned earlier) the polarity of the reference voltage to to the slide wire is reversed, point B of Fig. 2-1 will now appear as a maximum to the instrument, as point A did before it was located. This point will now be located in the same manner as has been described. When point B is located, the reference-voltage polarity is again reversed, the counter and register are reset and then released, and the instrument is allowed to look for the next peak. The negative peak does not produce data output.

Operation of the instrument begins when the start button is pushed. This sets the counter, register, control circuits, and reference voltage to conditions required to locate point A. After point A has been located and validated, the instrument generates command signals to transfer data to the printer and the multiplexer. The instrument then holds until signals are received denoting the completion of transfer. When these signals are received, the instrument, with the polarity of its input signal reversed, locates and validates point B in the same manner point A was located. Upon location of point B the input-signal polarity is again reversed, but no transfer commands are generated because B is a minimum point; instead, the instrument goes on to locate the next maximum. When all samples have run through the analyzer, the recorder registers a value less than 10 and the instrument automatically holds until new test samples have been loaded in the analyzer.

The printer is a simple unit that uses $\frac{1}{2}$-in. paper tape and is used to give the technician an indication of proper operation of the interface unit. Test results, accession number, and analyzer identification numbers are printed.

In order to provide sample identification, a three-digit accession number is typed into a small 64-word, 12-bit memory at the time a sample is inserted in the analyzer. The accession number is dumped out with the test result.

For stat tests the accession number is typed into memory at the address corresponding to the location selected for the specimen (generally 1–40) on the sample-holding turntable. There is a one-to-one relationship between the address in memory and the physical location of the sample on the turntable.

Other hardware devices, in particular block-transfer-control (BTC) units, are used to maintain the speed of data transfer between the external devices and the disk memory. At some times of the day there is high-speed transfer of the entire set of data contained in the satellite disk memory to the disk memory in the central computer, using the BTC and a coaxial-cable–video-line-driver system. "One"s are sent as positive pulses and "zero"s as negative pulses so that the system is self-clocking and requires

2. Automated Multiaccess System for Clinical Work 69

only one coaxial cable. Bit rates of 5 MHz are possible over the mile-long cable. By careful attention to the setting of a threshold detector at the receiving point, the data rate can be extended to 10 MHz.

B. System Operation and Software

Three basic sections form the computer system we are investigating for the collection, processing, and dissemination of clinical data. Each of them (the satellite computer, the central computer, and BASP) requires different software, although programs for the satellite computer can be tested on the central computer prior to being placed in day-to-day operation. In general, all programs for the satellite computer are compiled and debugged on the central computer, and a punched paper tape is taken to the satellite to enter the program on the disk. The original programs, such as the real-time monitor system, were compiled on a disk at the central computer and the disk was transferred to the satellite computer. All programs are stored on paper and magnetic tape so that a new disk can be generated if a disk failure occurs.

The satellite system operates essentially in a batch-processing mode, with interrupts designating the program to be called from the disk. There are three possible interrupts. One is to service a display console. The second services any device connected to the delay-line memory unit. The third interrupt, occurring twice a day (7:30 a.m. and 5:30 p.m.) transfers the data collected by the satellite to the central computer. Interrupts are not immediately responded to, but are serviced at the end of a program rather than at the end of a statement as in the usual priority interrupt system. Only short operations are done on the satellite system, and any interrupt is serviced in less than 5 sec, except for the two periods of the day when data are transferred between the two computer systems in about 15 sec.

A hardware hierarchy system services the consoles, the delay-line memory, and the central computer in that order. When a program has been completed, the satellite computer checks the I/O terminal for an interrupt. If no flag has been raised, the computer goes into a wait loop, which keeps checking the I/O unit and through it the two multiplexers. Raising a flag on any external unit then causes an immediate response.

A request for service causes the computer to read the first word from the I/O unit and bring in from disk the required subroutine, or to locate the subroutine if it is already in core. The subroutine reads in two more words, which for AutoAnalyzers are the accession number, the test result, and number of tests in the series. These data are then processed by the computer.

Interrupts are serviced on a first-come, first-served basis with the priorities assigned as above, and are serviced when the computer is free, not when an external device enters a request.

At the two times when an interrupt is received from the central computer, the data from the satellite disk are transferred in blocks to the disk unit of the central processor. The data transmitted at 7:30 a.m. consist of the test requests that had been entered into the satellite using the keyboard console in Clinical Pathology. From this information the central computer, operating as a batch processor with a FORTRAN IV program entered by cards, proceeds to generate and print out the work sheets and daily log for the technicians who will collect the samples and do the tests. At 5:30 p.m. all the information collected by the satellite during the day is once more transmitted to the central computer, where an appropriate program is entered and the required cumulative reports printed for the patient's records. The data on the disk are transferred to magnetic tape to provide a machine-readable record of that day's activity. These data along with data stored on tape from previous days are sorted to produce the cumulative reports.

Except for the hour used for clinical work, the central computer is free for batch processing of off-line data from patients and experimental animals. The batch processing can be interrupted on a cycle-stealing basis to dump data from BASP, in a burst mode, onto the disk for later retrieval and printout, plotting, or analysis. The central computer can also be operated in the usual foreground–background mode using priority interrupts.

To begin a day, the keyboard–display operator in Clinical Pathology turns on the computer and pushes "master clear," "single," and "start." This causes a disk bootstrap operation that loads the real-time monitor system and some operational subroutines into core. The monitor program initializes the computer.

Next a prepunched and labeled IBM card (Fig. 2-6) is inserted into the incremental card reader. The prepunched card, one of 50 cards punched for each patient at admission time, contains

(1) A six digit hospital number,
(2) A fiscal number for bookkeeping procedures,
(3) the patient's name,
(4) sex,
(5) date of birth,
(6) name of the attending physician.

The pressure-sensitive label affixed to the card is a carbon copy of the doctor's written order and may contain one or more test requests for the same department. Mixed-department tests (a chemistry and an x-ray

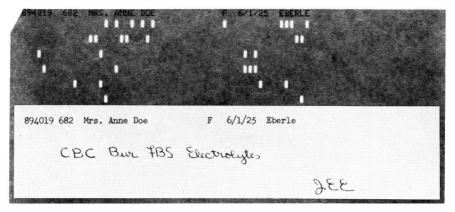

FIG. 2-6. Prepunched card for requesting service.

request, for example) are not allowed on the same request card. The incremental card reader sends the punched data from the card to the keyboard–display unit. After the card has been read and appears on the CRT display, it is ejected to be viewed by the operator.

While the labeled card is being read, the six-digit hospital number is stored in a buffer until the operator pushes the "display" pushbutton. This interrupts the computer, and a subroutine is called in to check the list of hospital numbers stored in the computer core memory. If the hospital number has not been previously entered, a light on the console is turned on.

Pressing the "display" button releases the typewriter keyboard, and the operator now enters the test requests listed on the card label into the memory of the keyboard–display unit. Any desired clear text may be entered at this time, but each test is preceded by an asterisk. A typical format appears as follows:

    894019 682 MRS. ANNE DOE F 6/1/25 EBERLE
    * CBC
    * BUN
    * FBS
    * ELECTROLYTES

When the data entry is complete and verified by viewing the label card alongside the information displayed on the console, the operator pushes the "transmit" button. The computer is interrupted, and a subroutine for servicing the entry of data is called in from the disk memory. A check is made against an ordered list of hospital numbers to see whether the desired hospital number is already in the computer. If the hospital number is not listed, the computer assigns an accession number, enters the data, and puts the proper dimensions after each test request before storing the page on disk.[3] This procedure of putting a card into the card reader and typing the test requests is continued until all the test requests have been entered through the keyboard–display console to the satellite computer. A listing of patients by wards is also entered when the cards are read, using the latest available information written on the test-request cards.

At this time the data on the disk of the satellite computer are transmitted to the disk of the central computer. All of the data are stored in a truncated ASCII code and are ready for print out.

From the data received the central computer (using a FORTRAN program) generates daily work sheets containing

(1) patient's accession number,
(2) patient's name,
(3) hospital number,
(4) ward,
(5) all tests requested for the patient from a particular department.

Three lists are printed. One (by wards) is given to the technicians who will pick up the samples. A second (by accession number) is the daily log of Clinical Pathology and lists all the tests to be done for that day. The third list (by sections in Clinical Pathology) is the daily log of each section.

Each technician is assigned a group of tests and receives a list on which to write the test results if the test is one not done on an AutoAnalyzer. AutoAnalyzer tests have direct computer entry, using the peak reader and delay-line memory system already described.

Whenever the satellite computer is not busy displaying or entering data, it accepts the next three words from the delay-line memory. If the system has been in operation for a period of time, several test results from a particular AutoAnalyzer will have been received and stored on the disk. As the AutoAnalyzers have a tendency to change calibration with time, calibrated samples are interleaved between patients' samples so that correction for analyzer drift can be made. Generally, ten standard samples of different concentrations are entered into the analyzer at the beginning of the day, and every fifth to eighth sample throughout the day is a standard.

The accession number 998 is assigned to all standards except for the

---

[3] Two switches had previously been set on the console to indicate to the computer that the page being displayed or entered was for Clinical Pathology tests. A different code number would have assigned the data to a page in another section of the disk memory reserved for a different department.

last sample in a run, which is always a standard and has the accession number 999. All patient samples have numbers less than 990.

Test results from AutoAnalyzers are stored on the disk in a series of three-word bins. Each analyzer has a number that protects a section on the disk for the storage of test results prior to a correction routine. When a series (generally 40 samples) has been completed, the standard number 999 is recognized by the computer and a drift-correction subroutine is brought into core along with the test results of the 40 samples. A correction factor is added to each patient's test, based on a linear interpolation of the change with time of two successive standards as read by the analyzer. The calculations are carried out in binary form by the computer and when complete are converted to ASCII and then inserted on the appropriate patients' pages.

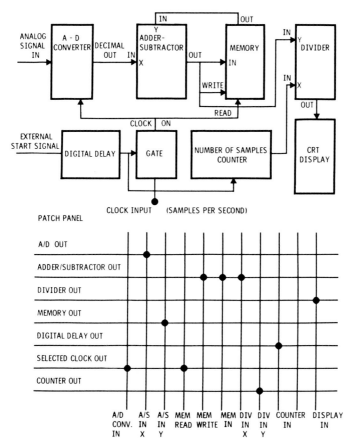

FIG. 2-7. Flow chart and control-panel pins for averaging on BASP unit.

At the end of the day the data collected by the satellite computer are transmitted to the central computer as described earlier. The FORTRAN program used in the morning is entered into the computer, and selected operations are performed by setting switches on the console. Data sheets identical to the morning run are printed out with the results inserted for comparison with the test results written by the technicians.

The test results are coded and written on magnetic tape along with the results of previous days' tests. This tape is then read into the computer and a cumulative report, from Monday through Sunday, is filled in for each patient and becomes part of the patient's record. Two tapes store all the patient data collected in a month and are placed in protective containers for long-term storage.

Software for the preprocessor BASP consists of the flow chart of an algorithm that is to be implemented. The program is set up by sticking pins into a panel at the junction points of the various arithmetic and memory modules indicated by the flow chart. As all channels are hardware-protected, any channel can be selected and a program punched up much in the manner of using an analog computer, but without the usual scaling problems. As an example, an averager flow chart and the implementation would look like Fig. 2-7.

The output of any memory can be displayed or transferred to the central computer by a pin in the control panel. Both display and long-term storage can of course be obtained simultaneously.

III. SYSTEM PERFORMANCE

The computer system described here is an experimental one in more than the usual sense. It is not an attempt to use standard equipment in a new or modified manner, but rather an attempt to apply engineering and programming capabilities to find the optimum hardware–software trade-off that will optimize performance and keep cost at a minimum. Therefore, a large amount of the system had to be constructed from microcircuit modules especially for instrument-interface devices and input-display devices that are adapted for clinical work.

A critical analysis of commercial units for remote data entry and display indicated that low-cost devices (Teletype machines) were noisy, unreliable, slow, and difficult to maintain, while keyboard–display units of the CRT type were too sophisticated, not easily adapted to the problems, and too expensive for the number of units that could be assimilated into a hospital environment.

To be financially practical the cost of time on any computer system assigned to the Clinical Pathology Department has to equal the money

2. Automated Multiaccess System for Clinical Work 75

recovered by a reduction of personnel or an increase in the number of tests that can be run in a given period by the same staff. As the number of patients is increasing and as a battery of tests done on all persons would improve the general health, it appears that increasing the number of tests that can be done, through automation, is the best route.

Considering the large amount of information available in a hospital and the necessary response time to a request for information, there is an optimum size and speed of storage for a system.

We have been doing trial runs of programs and procedures for one month at the time of this writing. Careful measurements have been made to see how close we have come to predicted capabilities and to check for malfunctions and design errors.

Most of the speed requirements are in the satellite system, which is operating on-line; the central computer system is generally operating in a batch-processing mode.

The most-time consuming program in the satellite unit is the one used to apply a correction factor to AutoAnalyzer test results by interpolating between standards interleaved with patient tests. This program reads in from disk the 40 completed tests and applies the necessary correction to each test result. Then the corrected value is converted to ASCII code and placed on the appropriate patient's page on disk. As random access to the disk is used, the time required is a maximum of 6 sec. In general, we find it to be less. Since the computer cannot be interrupted until the program is complete, and since not all programs take 6 sec, one can reasonably expect an average wait of 3 sec before a display unit is serviced.

The actual time required to service a display unit is 0.2 sec, but if ten display devices requested service at the same time, the tenth one could have a wait of as much as 6 sec for the AutoAnalyzer program, plus 2 sec for the display units.

Some disk memories proposed by other computer manufacturers had access times of 0.5 sec and 500 K words of storage, or access time of 16 msec and 300 K words of storage. We chose a disk of 150-msec access time and 1.5-million words of storage. The access time is adequate, as 40 disk accesses (that is, one run of an AutoAnalyzer) is probably the maximum that will be done in any one program and the central processor time for the program is negligible. If necessary, the program could be split into two parts and the display units serviced in 3 sec at the end of the first part. Because of the priority system, all display units that had requested service would be satisfied before the second half of the analyzer program was run.

Each display or keyboard display unit has its own delay-line memory, and therefore once data are received or transmitted the unit disconnects from the computer. The time for the transfer of a page from or to the core memory of the computer is 16 msec, which is the delay time of the display memory. To expedite the transfer of patients' pages from the disk to core memory and to the display unit, all pages are put on disk in ASCII code and are in core image.

At present we have 250 in-patients, but expect to expand to a 500-bed hospital shortly. Based on 500 in-patients, we have split the satellite disk into the slices shown in the following tabulation. The words assigned to each department are for one day's data.

SUBDIVISION OF THE DISK (CAPACITY 1.5-MILLION WORDS)

Department	Maximum number of pages	Number of characters per page	Number of words
Disk slices for patient pages by departments			
Clinical pathology	500	1600	400,000
Pharmacy	500	300	75,000
Pathology	60	400	24,000
Medical diagnosis	60	800	48,000
ECG	20	400	4,000
EEG	20	400	8,000
Radiology	40	400	8,000
		Total	567,000

Disk slices for automated devices bins	
Maximum number of bins	100
Number of words reserved per bin	192
Total number of words	19,200
Disk slice for subroutines	100,000 words
Disk slice for subroutine buffers	200,000 words
Disk slice for storing all AutoAnalyzer data	10,000 words
Approximate total	896,000 words

As only a little more than half of the disk is reserved, we have adequate space to put out-patient data as well as in-patient data on the disk for later transfer to digital tape for long-term storage and statistical analysis. At 20 lines of 80 characters each, the Clinical Pathology page can display 20 test results, one per line. Generally, a maximum of 8 tests are done on a patient in a day, and other display lines are used for comments by the pathologist, which are read by the physician using a CRT display unit. Pages assigned to other clinical departments are for clear-text information that is to be transmitted to the physician as soon as required and considerably before the results get into the patient's chart. Coded data, using standard codes (such as in the International Classification of Disease, PHS publication 719) are stored for later retrieval and correlation analysis. Our present 2800 tests a day for Clinical Pathology could expand to 10,000 before the disk was loaded. We expect eventually that combined tests on in-patients and out-patients will reach 7000 per day.

In order to handle the increased number of tests, many of which will be done by AutoAnalyzers, more channels will have to be made available. To add a channel to the number 1 or number 2 multiplexer requires the addition of ten microcircuit modules. The multiplexer–delay-line-memory has an almost unlimited expansion capability and in its present form can service 20 autoanalyzers and as many numerical data-input devices as desired, up to a total of 64 external units. Multiplexer number 2 can support 64 display or keyboard–display units. We envisage 20 display devices as our probable maximum requirements.

Reliability in any electromechanical system depends in large measure on the quality of the components used to construct the system. One has to depend on the reputation of the supplier of the central processor, which is mostly electronic, and the suppliers of peripheral units, which are mostly mechanical devices. Other than from the larger computer manufacturers, a computer system such as described here is generally assembled by a manufacturer who makes the central processor unit and purchases the peripheral units from other manufacturers. The peripheral units available cover a wide range of performance, reliability, and price. Some manufacturers stress service as a prime asset for their proposed systems. At best, however, service is generally available only after a delay of several hours, and in some cases a day or more.

In an on-line system a failure at any time for two hours generally represents a loss of that day's computer operation. Because of this, we have selected the units of the computer system to have the following characteristics:

(1) A mean time before failure (MTBF) as long as possible (generally 500 hours to several years).
(2) Electromechanical design to facilitate easy servicing by an in-house service group.
(3) The simplest design and minimum number of units to do the required on-line job.
(4) Compatibility with other units to provide redundancy.

To meet these requirements we surveyed the computer journals for reliability tests on peripheral devices. Each manufacturer was requested to supply MTBF data from previous experience or data calculated on the

CPU and peripheral units. Surprisingly, the larger computer manufacturers did not supply such data.

We have an engineering group in house who can service the computer system, and have assigned personnel to regular preventive maintenance. As the number of different modules used in the system is not large, a selected group of spare parts is kept on hand. During a shakedown period of two months we had failures in three modules. Each failure was traced and located in less than 15 minutes. In no case was the system non-operational, as the failures were in peripheral devices of the central system and not in the satellite system.

Our primary aim is to keep the satellite hospital system, which will eventually operate unattended 24 hours a day, in good working condition. A minimum system of a CPU and a disk memory is used. The disk memory, a CDC 9342, is considered well designed by the computer industry, and with proper preventive maintenance should have a predictable service life. There are new disks available with fewer mechanical parts, and these should be even more reliable. Fixed-head disks or drums, although of less word-storage capabilities, can be used if further improvement in reliability is required. We have an advantage in that the central computer system has parts and units that are interchangeable with those of the satellite unit. In effect, there is a 100% redundancy on equipment that must operate. Our past experience indicates that the special units, such as the multiplexer–memory system we have constructed, will operate for several years without failure.

The two Teletypes can be used as printers of work sheets if a failure occurs in the line printer at the beginning of the day. If the central computer is down at the time the satellite is to transfer its data for long-term storage, a spare disk can be put on the satellite and the disk for that day saved until the central computer is repaired.

All transactions to and from the satellite computer could be recorded on magnetic tape and completed off-line, except for the display of data during the day. We are not using this procedure at present until more data on reliability can be obtained by running the system.

In general, procedures for handling the most probable system breakdowns have been established, and if the computer system is completely out of operation there is enough paper information at all times during the day to permit the completion of clinical tests. Once the computer is repaired, the most recent data can be entered through the keyboard–display units.

All keyboard–display units can, by a switch, be made applicable to any department for data entry, and eventually a spare unit will be made available to replace a defective unit until it is repaired.

IV. CURRENT CLINICAL EXPERIENCE

Considering the short period we have had the computer system and that we will not be operational on a hospital-wide basis for five months, we cannot report much clinical experience. However, a few pertinent facts have come to our attention.

The programs employed to generate daily work sheets, logs, and cumulative reports were rewritten several times before the format was acceptable to the pathologists. We found that some test results required two decimal places (e.g., pH = 7.06), others were to be written with one place (e.g., K = 5.3 me/l), and in some cases a decimal place was not allowed at all (e.g., Na = 143 me/l).

We found that a display unit with a 30 × 40 cm screen was too large and that characters about 1.5 times as large as typewritten characters, 4 × 3 mm, were preferred instead of the 7 × 5-mm characters we provided.

Clinical departments insist on keeping the original requests for service, yet the business office requires a copy for making out the patient's bill. We found that in most cases the department supplying the service calculated the charges and entered them on a special charge card for the business office. As we wanted a simple system with the least amount of transcription of information, we decided to use the data obtained from the original request card (with the doctor's order on it) and the pricing information entered at a keyboard–display unit at the completion of a test to generate the charge cards for the business office computer.

The device used to pick the AutoAnalyzer peak and dump the test result into the computer has worked well except for one situation we had not anticipated: sometimes a technician will offset the analyzer recorder so that small pertubations that are actually below zero (that is, negative peaks) are placed on the positive scale. The unit we had devised interpreted these peaks as being legitimate. We have since inserted a circuit that inhibits the readings of all peaks less than 5 on a 100-graduation scale. Also, our original concept was to look for a peak every 15 sec, but we found that technicians had a habit of holding the turntable for a period of time while a new test was put in place, such as for a stat test. The device we finally constructed works as previously described and is independent of time.

V. CRITIQUE

As noted earlier the business office and other sections of the hospital, in particular the admissions office and the out-patient clinics, have a major influence on the automation of the clinical laboratories. This situation exists because there has not been a universal procedure for admitting a patient and requesting a service such as a clinical pathology test, an x-ray, or a drug. These departments and others have had their own procedures for test requests and for billing. A multiplicity of methods and forms for requesting service and reporting results can only increase the personnel time required to learn and handle the many types of paper work, and it makes any real improvement by the installation of a computer a very difficult project.

Our investigation of the whole hospital system enabled us to propose a single procedure for handling all service requests that requires a minimum of learning and handling by personnel and that is easily communicated to the satellite computer. This procedure, described earlier, makes use of a prepunched IBM card with a press-on label, which is a carbon copy of the doctor's order, on the face of the card. A trial operation of a card reader constructed by us showed the feasibility of the procedure. There are at present units made by Hewlett Packard and Motorola that may be adaptable to reading the card. The difficulty in both of these units is making the card visible to the operator after it is read into the display unit, so she can enter the written doctor's order through the keyboard and then see the CRT display and the card side by side for verification before the data are transmitted to the computer. We are continuing to improve the card reader we have constructed and are working with the manufacturers to see if a modified commercial unit can be obtained.

The system presently used for marking all reports, service requests, and charge cards employs an embossed card, made at admission time, and an imprinter at each of the wards. All order sheets, charge cards, and service requests are imprinted. Originally, the imprinted information on charge cards—fiscal number, hospital number, name, etc.—was scanned by an optical reader at the business-office computer. However, the imprinting was often improperly done (being illegible or crooked) and the optical reader rejected a large percentage of the cards. The optical reader has been removed and replaced by keypunch operators who read the imprinting and punch the information into the card.

Our concept of an improved procedure for handling service requests requires the following at or before admission time: the punching of 50 IBM cards with the information described earlier, the printing of 20 doctor's-order sheets with space for ten series of tests on each sheet, and the printing of 100 pressure-sensitive labels to be used for identifying test tubes and other samples, the patient's charts, patient's room, nurses' sheets, and all of the various pieces of paper now being imprinted with the embossed card.

In order to automate the generation of the cards, the doctor's-order

2. Automated Multiaccess System for Clinical Work 81

sheets, and the labels, we have ordered an IBM 526 keypunch and two Kleinschmidt model 311 printers. These units and a small memory–control system of our construction will be placed in the Admissions Office. While the admissions personnel are typing up the admission form, selected information will be stored in the memory–control unit. On command, the keypunch will generate the 50 cards, the first printer will print the same information at eleven places on each of the twenty doctor's-order pages, and the second printer will print the data on the 100 labels, which are arranged two across on an $8\frac{1}{2} \times 11$-in. sheet. This automation is necessary so that additional personnel are not required and, more important, so that consistent information is on all papers pertaining to a given patient. A packet of cards, order sheets, and labels will go with each patient at admission, and further copies will be made at the admitting office upon request by a ward secretary or nurse. The labeled cards eventually are retained by the department from whom service is requested.

Because our business-office computer center uses programs supplied by the computer manufacturers and finds it difficult to modify them to accept paper-tape or magnetic-tape input at this time, we are modifying a keypunch to generate the necessary charge cards from data transferred to the central computer at the end of each day. A card with the information printed on the top is necessary because the business-office computer may reject a card that contains erroneous information due to an improper code or incorrect format; the printing enables personnel to read and correct such cards. As this procedure of punching cards will require four hours of off-line operation, we hope that when a more modern computer is installed in the business office, a much faster input will be allowed. However, the procedure presently being set up will replace two or more keypunch operators now used to punch and verify the charge cards.

Our central computer system could, of course, be used to do the business-office operation if necessary. However, this type of operation is not in keeping with our NIH grant for this developmental program. We will use the time available on the central computer system for patient-monitoring and research purposes, as we are a Medical School with an attached teaching hospital.

The development of a keyboard–display and a display-only unit was undertaken in an attempt to reduce the cost of these devices and adapt them more closely to the way we envisage them in a hospital system. We are not satisfied with the 800 characters we display now and are trying to upgrade the unit to 1600 characters and also reduce the flicker. The quality of the characters is excellent and allows a physician to scan a page without the mental translation of letters into words that we have noted on other commercial display units. A major effort is being made to produce a high-quality display of 1600 characters at a parts cost of $1000 so that the unit can be manufactured for $2000 in quantity and be economically practical for small hospitals of 100–200 beds.

At present we are considering using one of the peak-picker–digitizer units for each AutoAnalyzer channel. However, these units cost about $500 each, and in an effort to keep costs down an analysis is being done to determine whether one peak-picker–digitizer unit can be multiplexed for any number of autoanalyzers by the addition of registers. To let the computer pick the peaks on 20 AutoAnalyzer channels would, considering that peaks on several units could occur simultaneously and require interpolation of the standards and tests, make it impossible for the computer to collect all of the peaks and yet provide a fast response to a display request.

To lower the work load on the keyboard–display operator in Clinical Pathology and also increase the accuracy of reporting test results, we are developing a simple keyboard that can enter data directly from a test station, such as at a pH meter, to the multiplexer–memory unit. The technician would key in the result instead of writing it in a log. One unit will be tested at first, as we are concerned with the amount of time the technician may spend entering the data. Much of the information, such as type of test, would remain in the unit and not have to be keyed in each time a test result was entered.

By the time this book is published, we will be in full operation. The delivery of card punches and printers and our modifications to make them useful will require five months. During this time, improved card-reader and display units will be developed and, as we are an engineering group with our own printed-circuit facility, duplicated as desired.

VI. EVALUATION

The satellite computer as presently organized can accept all of the information generated by Clinical Pathology, Pathology, Radiology, ECG, EEG, Medical Diagnosis, and Pharmacy, store these data, and have them available for CRT display, using only three hours of computer time each day. Almost half of a disk memory is available, so that additional data can be collected and disseminated. At most, 5000 words of memory are being used, and several more fast-access programs could be added.

Sorting and printing of daily logs, work sheets, and cumulative reports takes an hour on a 300-lpm printer. This is much less than the ten hours that would be required if a Teletype or one of the relatively inexpensive page printers were used. Yet it is impractical to use a line printer for only one hour a day and a tape deck for only two or three minutes a day.

Several systems for clinical work are using the small tape units from

2. Automated Multiaccess System for Clinical Work

Digital Equipment Corporation. However, as we desired a 5-sec display response to an inquiry, we had to use disk memory.

The above are several of the reasons we decided on a satellite and central-computer system concept. Further, we were able to purchase the combined system for less than systems that operate in a foreground–background interrupt mode with their expensive supervisor system. Also, as noted earlier, the usual systems would require a large amount of programming effort and use almost all of the computer time to do just the clinical areas of our hospital. Time-shared systems operating with priority interrupts cannot do what is required for on-line clinical work at any reasonable cost, and business type computers are completely unsatisfactory. At present we have in an eight hour day five hours of satellite time and seven hours of central computer time available for work outside the clinical service areas.

The satellite system with a CPU, disk memory, Teletype, multiplexer-delay-line memory, two display units and one keyboard–display unit, and an interface to 12 AutoAnalyzers cost $77,000. We estimate that a total system including the devices to generate the request cards and labels at admission time and ten more display units will cost $120,000. As this equipment will handle the on-line operations of a hospital and replace at least two keypunch operators and two laboratory technicians, it should pay for itself in five years, during which time a large increase in clinical data can be handled without additional cost. The central computer system cost $160,000.

The satellite system can be used with any third-generation computer system for the printing and long-term storage operations. We use a direct transfer of data between the two systems, but the same results could be obtained using a tape unit at the satellite or possibly by a physical transfer of the satellite disk pack to a central computer. Most third-generation computers will accept data on a cycle-stealing basis, and this procedure can also be used. A major advantage of having both the satellite and the central computer constructed of identical modules is the redundancy this provides.

Our analog preprocessor BASP is relatively inexpensive and has a cost dependent upon the number of channels and data points per channel. A four-channel unit with 1024 words per channel operating at a maximum sampling rate of 250,000 words per second per channel would cost about $30,000. The addition of four more channels would cost another $10,000. These are high-speed channels useful for on-line analysis where data rates of 100,000 samples per second and up to 10-million arithmetic calculations per second are required.

The satellite computer, by having the delay-line memory and the two multiplexers, is capable of operating as a batch processor that is independent of the data-flow rate from the various input devices. The addition of a real-time clock and another multiplexer–delay-line-memory unit will probably give the system the capability of monitoring slower signals such as ECG, EEG, respiration, blood pressure, etc. and of typing out required messages. We have not fully investigated this possibility as yet, but our work on the time-shared operation of BASP indicates the usefulness of this approach. We naturally want to load the satellite computer so that it is doing something useful 24 hours a day so as to reduce its effective cost per operation.

We are convinced that a small computer, properly oriented and dedicated to the on-line operations in a hospital, is the most practical system in terms of response time and cost. Therefore we expect that future hospital data processing of the "hospital information systems" type will use the above principles in preference to some time-shared operations that cannot do the necessary clinical work and the bookkeeping in an on-line operation. Those large systems that could possibly supply the necessary service are prohibitively expensive and in fact are multiprocessors working in the satellite–central computer mode described here.

The number of patients on which analyses, at a high or low data rate, can be done on the preprocessor BASP is determined by the amount of memory available. All of the high-speed channels in the processor will not be in use at the same time, and if, for example, one of ten channels operating at a sampling rate of 100,000 data points per second were free, it could be divided into 100 channels operating at a sampling rate of 1000 data points per second. This rate, of course, is adequate for many biological signals. Therefore we have begun interfacing a rapid-access drum memory to BASP to determine whether we can monitor 1000 patients' signals at a sampling rate of 1000 data points per second and do some simple operations such as measuring time intervals, averaging, auto- and cross-correlation, algebraic calculations such as used for cardiac output, and any other procedures that do not require several hundred to several thousand program statements as is common in the usual on-line digital computer operation.

We are finding that the programming procedure for BASP, when used for physiological research on animals and patients, is intuitive for physiologists and clinicians who are applying sophisticated mathematics to the biological signals. Further, when BASP is operating in the time-shared mode all channels are available at any time and are hardware-independent. Therefore another channel can be placed in operation without delay and without in any way interfering with other channels. The program, of course, is alterable during an experiment. Thus the investigator need not be concerned with supervisor programs and priority interrupts as in normal systems. In this sense the processor appears more useful than the usual small computer systems, which are programmed in assembly language (or in some cases in FORTRAN) and which require the investigator to develop expertise in a generally unfamiliar area.

We feel that there will be a place for this type of device, especially where multichannel operations are required, such as for doing five averages and five variance analyses of multielectrode EEGs where the cost of ten on-line devices (which, because they are fixed programmed, cannot be used for other types of analyses) would far exceed the cost of a unit such as BASP. The addition of the drum memory and a procedure in which the results of a previous time-shared channel are being placed on the drum while the next channels, data, and program are being sent to the processor, will permit a greatly expanded number of channels at lower cost than core memory.

Cold hard facts that point "the" way to improved patient care through automation and computer usage are unfortunately not available to us at this time. We have presented a concept and some of the pitfalls one finds in a hospital environment and hope that a better-defined statement of what needs to be done and some solutions may have evolved from our investigation.

Reference

Mueller, William J., Paul E. Buckthal, Philippos Lambrinidis, Karl E. Schultz, and Leo F. Walsh (1968). BASP—A Biomedical Analog Signal Processor, *Proc. Spring Joint Computer Conference*, Vol. 32, pp. 151–159.

CHAPTER 3

An On-Line Graphic Inquiry System for Analysis of Tumor Registry Data

HARVEY S. FREY

DEPARTMENT OF RADIOLOGY, UNIVERSITY OF CALIFORNIA, LOS ANGELES, CALIFORNIA

I.	Introduction	87
	A. The Problem	87
	B. A Solution	88
	C. An Example	88
II.	Implementation	96
	A. Hardware	96
	B. Data Organization	96
	C. Software	98
III.	Performance	102
IV.	Critique	103
V.	Discussion and Another Solution	104
	References	105

I. INTRODUCTION

A. The Problem

Tumor registry data are more often collected than consulted. The demands on time, personnel, and patience required to search, subclassify, and analyze a large file are such that the busy physician is rarely able to take advantage of the information so laboriously collected. Many files maintained on punch cards are designed more for analysis by sorting machine than by computer. A file kept on magnetic tape relieves more of the tedium of sorting, but is still far from ideal. Such a file must be searched sequentially. Searches are usually run in "batch" mode and if improperly specified may require several runs before the desired information is found. This problem is not peculiar to tumor registries, but is found in almost identical form in the areas of clinical trials, research bibliographies, and many hospital records.

B. A Solution

Several recent developments in computer technology have made a more convenient approach possible. These are (1) direct-access mass storage devices and the ability to use them in the context of a high-level language, (2) time sharing, and (3) cathode-ray-tube display units connected to the computer that allow the user to interact with the central processor.

The author has taken advantage of these developments to write a program for analysis and retrieval of tumor registry data. It was felt that such a program could have several advantages over the more conventional approaches. The user would have immediate access to any part of a very large file and could easily specify any portion or portions of it for deeper analysis. By using the rapid and responsive display console, he could bring the full power of the computer to bear on any groups so selected to test hypotheses or characterize data. The results of such computations could be immediately displayed graphically so that the user could refine his hypotheses, change his search requests, and, by intimately interacting with the data and the machine, find the answer to his real question. The final question may not necessarily be the one he asked first.

C. An Example

To give an impression of how the program works, let us follow an hypothetical investigator who, for example, is interested in squamous cell carcinoma of the tongue, as he uses this program. He loads the program and seats himself at the display console, which is somewhat like a large television screen. A list of variables, such as site, histological type, etc., is displayed on the screen along with the control words ALL, PROCEED, and START OVER (Fig. 3-1). He detects on the word SITE with a light pen.[1] The display disappears and is replaced by a list of organ systems. For example, BUCCAL CAVITY & PHARYNX, RESPIRATORY SYSTEM, etc. (Fig. 3-2). He detects on BUCCAL CAVITY and a list of sites within the buccal cavity appears (Fig. 3-3). Detecting on TONGUE he obtains a display that says BASE OF TONGUE, OTHER SPECIFIED PARTS (Fig. 3-4). Since he wants both he detects on the word ALL. He now returns to the main display by detecting on PROCEED. This time he

---

[1] That is, he touches the word on the screen with the tip of the pen and presses a pedal.

3. An On-Line Graphic Inquiry System

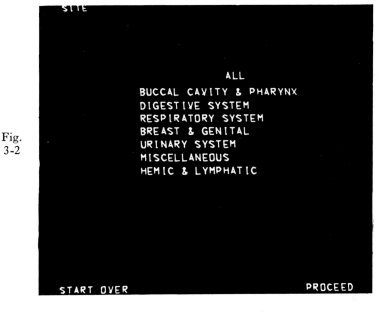

Fig. 3-1

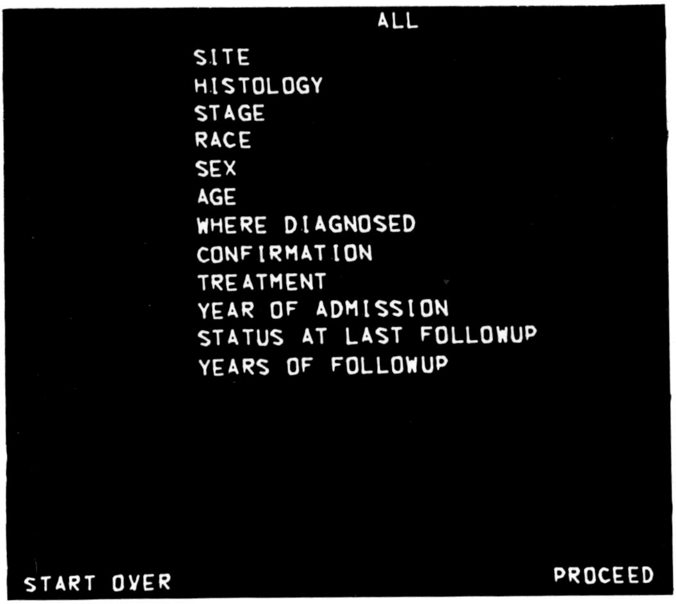

Fig. 3-2

Fig. 3-1 (top). Scope display initiating data-retrieval program.
Fig. 3-2 (bottom). Second display (organ systems) in program sequence.

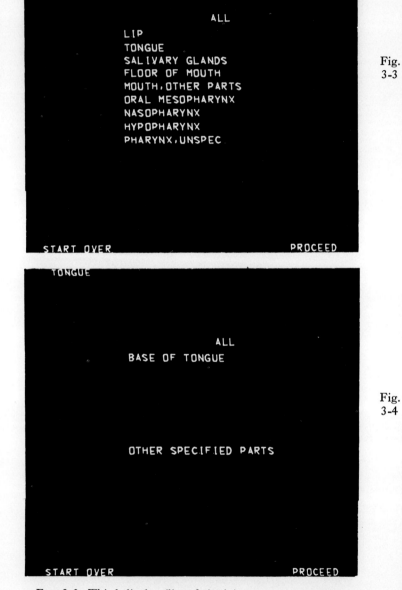

Fig. 3-3.
Fig. 3-4.

FIG. 3-3. Third display (list of sites) in program sequence.
FIG. 3-4. Fourth display (localization) in program sequence.

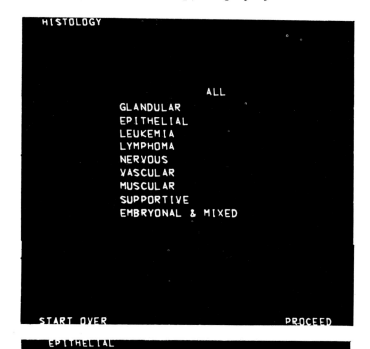

Fig. 3-5

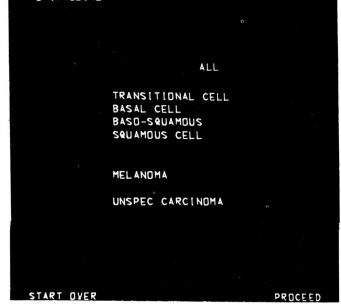

Fig. 3-6

Fig. 3-5. Fifth display (first-level histology) in program sequence.
Fig. 3-6. Sixth display (cell type) in program sequence.

detects HISTOLOGY, and the resulting list says GLANDULAR, EPITHELIAL, etc. (Fig. 3-5). A detect on EPITHELIAL gives a display that lists TRANSITIONAL CELL, BASAL CELL, etc. (Fig. 3-6). After a detect on SQUAMOUS CELL, the display does not change, signifying that this variable is not further subdivided. A detect on PROCEED returns to the main display, and another such detect puts on the screen a list of those variables that have been selected for the search (Fig. 3-7). The light pen is touched to this list and immediately names and

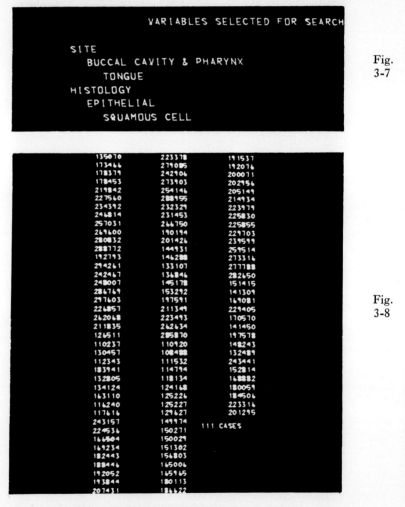

Fig. 3-7. Display initiating a data search.

Fig. 3-8. A typical display of hospital numbers obtained in a search.

3. An On-Line Graphic Inquiry System

hospital numbers of selected cases begin to appear on the screen (Fig. 3-8). (For purposes of publication the names have been suppressed). These are all of the cases in the registry that have squamous cell carcinoma of the tongue. Detecting on this list erases it and replaces it with a list of options (Fig. 3-9). One option causes all the information on each case to be

```
PRINT OUT CASES
DISPLAY NAMES
ACTUARIAL ANALYSIS
SUMMARY BY VARIABLE
START OVER
```

FIG. 3-9. Option-selection display following data search.

printed out on the line printer for later reference. Another option, SUMMARY BY VARIABLE, causes the main display to reappear, but this time when the detect is made, next to each number on the list is displayed the number of cases which fall in that category. For example, if

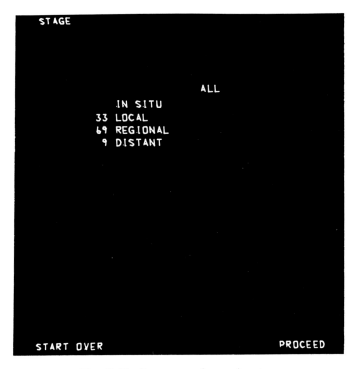

FIG. 3-10. Summary of cases by stage.

STAGE is detected, a list of stages is displayed with the number of patients falling in each category (Fig. 3-10).

The option ACTUARIAL ANALYSIS (Fig. 3-9) allows up to five subgroups to be selected. The selection is done in the same manner as the original selection. For example, group 1 might be cases with regional

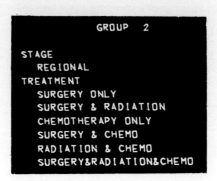

Fig. 3-11

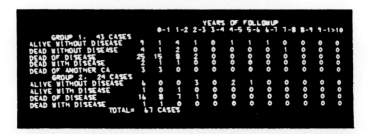

Fig. 3-12

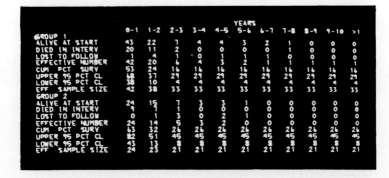

Fig. 3-13

FIG. 3-11. Display initiating actuarial analysis.
FIG. 3-12. Status table generated in actuarial analysis.
FIG. 3-13. Survival table generated in actuarial analysis.

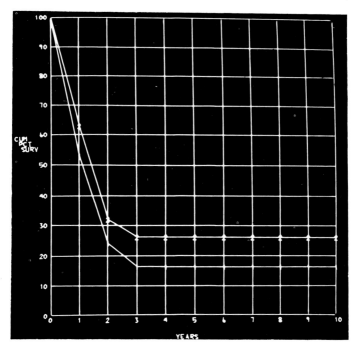

FIG. 3-14. Graphic display of survival values.

disease treated by radiation, and group 2 cases with regional disease treated by other means (Fig. 3-11). When the subgroups have been selected, another detect causes the display of a table containing the status at last follow-up and length of follow-up for each group (Fig. 3-12). A detect erases this table and displays a survival table for each group using the standard Berkson and Gage algorithm (Fig. 3-13). Another detect causes a graph of the survival curves to appear (Fig. 3-14).

At any point in the program pushing a certain function key on the console will cause whatever table, graph or list is on the screen to be printed out for permanent reference.

The options may be reentered as many times as desired so the data can be analyzed in many ways. As can be seen, this presents a rapid, convenient way to interrogate a large file and "browse" through the data, reformulating the search strategy as the data present themselves.

The program is applicable to files with fixed format, numerically coded fields, and fixed-length records, and which can be updated in batch mode. It is not designed for retrieval of information on a particular patient, but rather for statistical studies of patient populations. A large number of existing files doubtless fall into this category. The more general retrieval problem will be considered in a later section.

II. IMPLEMENTATION

A. Hardware

The program has run successfully on an IBM 360/40 with 128 K bytes of core, on an IBM 360/75 in a fixed partition of 100 K bytes, and on the 360/75 in a time-sharing environment ("multiprogramming with variable number of tasks"). The display unit has been an IBM 2250 Model 1 and more recently a Model 2, using 4 K bytes of a 2840 Model 1 Buffer. The data sets have resided first on IBM 1316 disc packs and currently on 2316 disc packs on a 2314 drive. Since all these systems are supposed to be compatible, each conversion required little more than a bottle of aspirin.

B. Data Organization

The program uses two direct-access data sets and a temporary file.

1. THE DATA FILE. The data file contains the tumor-registry records. Each record corresponds to one tumor of one patient, and contains 48 bytes of data. There are currently approximately 10,000 such records. All cancer cases known to the hospital are coded on work sheets. These are sent to the central facility of the California Tumor Registry, where they are checked and punched on standard IBM cards (80 bytes). New cards are returned to the hospital yearly. These cards, reformatted and compressed to 48 bytes, are the data base. The format of each record on the disc is shown in Table I.

The site codes are based on the International Statistical Classification of Neoplasms (International Classification of Diseases, 7th Revision, World Health Organization, 1955), as used by the California Tumor Registry. The digits of the code have been separated to provide three hierarchies of site. The codes for organ systems are derived from the first two digits of the ISC code by subtracting 13, giving a range of 1 to 7. "Organ" corresponds to the third digit of the ISC code, and "organ part" to the first digit after the decimal point, with alphabetic characters removed.

The histology codes are modified from the American Cancer Society's Manual of Tumor Nomenclature. Two hierarchical levels are used. The other variables have only one hierarchy. The data set is sorted by organ system and organ. The other fields are not sorted upon, so cases are randomly distributed with respect to them. The file is recreated at yearly intervals, when new cards are received.

3. An On-Line Graphic Inquiry System

TABLE I
Format of 48-Byte Record on Disc

Field	Bytes	Contents	Range
1	1	Organ system	1–7
2	2–3	Organ	1–9
3	4–5	Part of organ	1–9
4	6–7	Histology, major	1–9
5	8–9	Histology, minor	1–9
6	10	Stage	1–4
7	11	Race	1–6
8	12	Sex	1–2
9	13	Age	1–9
10	14	Where diagnosed	1–7
11	15	Confirmation	1–4
12	16	Treatment	1–8
13	17–18	Date of diagnosis: month	1–12
14	19–20	Date of diagnosis: year	0–99
15	21–22	Date of admission: month	1–12
16	23–24	Date of admission: year	0–99
17	25–26	Follow-up: months	1–12
18	27–28	Follow-up: years	0–99
19	29	Status at last follow-up	1-9
20	30–34	Standard nomenclature, site code	0–99999
21	35–42	Name	Alphabetic
22	43–48	Hospital number	0–99999

2. Control File. The control file contains control information, labels, and pointers. It consists of 325 80-byte card-image records. The first 9 records contain an array of pointers to the data file for each system and organ. Thus, when a particular site has been selected, the search starts at the appropriate place in the data file. Record 10, when read into the program, becomes an array that gives for each system the number of sites it contains. The next 10 records contain alphameric information used in the program for labels and titles.

Starting at record 21 is a table that describes the variables and their hierarchy structure. The first four fields tell where the given label belongs in the structure, e.g., if the numbers are 1261 this means that this record has the label for the first level of the sixth level of the second level of the first variable. 1260 means the sixth level of the second level of the first variable. 1200 means the second level of the first variable, and 1000 means the first variable.

The next field contains the alphabetic name of the level. It is used to form the display printouts. If this is not a terminal level, the next two fields point to the first and last records of the next lower level of names in the control file. For example, the first record is 1 0 0 0 SITE 33 39. This means that the first variable is SITE and its levels will be found from records 33 through 39 inclusive, in the control file. The second level of SITE is DIGESTIVE SYSTEM so record 34 is 1 2 0 0 DIGESTIVE SYSTEM 125 133. Record 130 is 1 2 6 0 BILE PASSAGES AND LIVER 249 250. This is the sixth level of DIGESTIVE SYSTEM. Record 249 is 1 2 6 1 LIVER. This is the first level of BILE PASSAGES AND LIVER. Since there are no pointers, this is not further subdivided. Thus the names and hierarchical interrelationships of the variables can be easily changed, and the program applied to an entirely different application simply by using a different control file.

3. TEMPORARY FILE. The temporary file is a sequential data set for temporary storage. When the major file is searched, the cases selected from it are placed in the temporary file and all subsequent analysis is done on this smaller number of records.

C. Software

1. LOGIC OF SELECTION AND SEARCH. The program assumes that the major variables are logically independent and that the values of a given level of a variable are mutually exclusive. This reflects the philosophy usually associated with fixed-format numerically coded files. For example, a person with cancer of a given site could conceivably be of any age or have received any type of treatment. Naturally there will be null sets—for instance, males with carcinoma of the cervix—but this does not effect the overall logic.

The logical operations most natural to this type of structure are intersection (AND) between variables and union (OR) between levels of a variable. The hierarchical structure renders each list sufficiently short that negation, if desired, can be achieved by affirming all members at that level except the ones not desired.

The program assumes at the start of the search that the investigator is interested in everything. If a variable is not selected with the light pen, then no value of that variable or any of its sublevels will cause a case to be rejected. Selection of a variable means that the user desires to restrict its range. Once a variable has been selected, the program assumes that none of its possible values is wanted. The user specifies those that *are* wanted by selecting them with a light pen, thus OR'ING them together. By detecting ALL he may specify that all are desired. The selections are stored in four arrays of flags, one for each level of hierarchy, the highest level being the variable names themselves. These flags are initialized to zero if the corresponding level has a lower level and to 2 if it is a terminal level. If a list item is selected with the light pen, a zero is changed to 1, or a 2 is changed to 3. If ALL is detected, the next higher flag is changed to 3 for a level or zero for a variable. Thus for a variable, zero means "ignore its values" and 1 means "consider its values." For levels of a variable, zero or 2 mean reject records with this value, 3 means do not reject records with this value, and 1 means look at the next lower level of flags in order to decide. As noted above, the variable SITE occupies a special position. It was assumed this was the variable least likely to be unspecified, so it was used as the basis of ordering the main file. If SITE is not selected, the entire file will be read and every record tested. If SITE is selected, the flags for each organ system are considered in turn. The logic is charted in Fig. 3-15. The logic used to accept or reject a case after it has been read in is diagrammed in Fig. 3-16.

2. THE PROGRAM. The program is written in FORTRAN IV. It was originally written in the E-level subset and bears many of the scars of its evolution, e.g., arrays are limited to three dimensions, and much alphameric information that might have been defined in data statements is instead read in from the control data set.

3. GRAF. The graphics is performed with GRAF (Graphic Addition to FORTRAN), a group of subroutines originally written by Joseph Yeaton and Alex Hurwitz at our facility. The GRAF language offers a simple and convenient way to use the IBM 2250 via FORTRAN. Since the graphics is the only portion of the program that might be unfamiliar to a good FORTRAN programmer, it probably deserves elaboration.

The author knows of several projects in which graphics was not used because the person in charge felt that it was very difficult or outre. Some of the industry-supplied languages contribute to this impression. To demonstrate the basic ease and simplicity a well-designed language can impart, let us consider GRAF briefly.

The central concept of GRAF is that of a display variable. This is a pointer to a list of orders that is to be sent to the display unit. The space for these lists is automatically obtained and released as required, freeing the programmer of much bookkeeping. A variable is declared to be a display variable by a call to DISPLA. For example,
  CALL DISPLA (A,B,C).
Orders may be placed in the display variable by calls such as
  CALL PLACE (A,100,2000)
  CALL LINE (B,3.2,7.8)

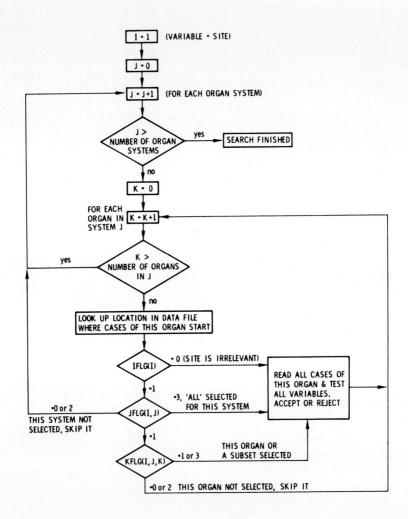

Fig. 3-15. Flow diagram for selection software.

3. An On-Line Graphic Inquiry System

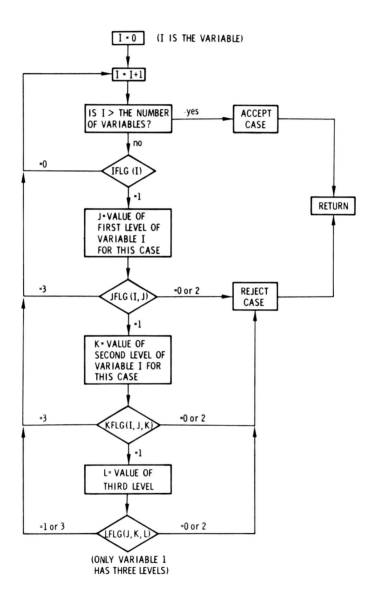

FIG. 3-16. Flow diagram for acceptance–rejection decision software.

        CALL POINT (C,20,50)
        CALL CHAR (A,'WORD', 4,69)
Character information may also be put up with the FORTRAN write statement. The characters are converted and placed into a dummy buffer, and a call to WRFMT$ transfers them to the display variable. For example,
        WRITE (4,200) X,Y
    200 FORMAT ('1X=', F5.3, 5X, 'Y=', F5.3)
        CALL WRFMT$ (B)
Orders in a display variable are transferred to the screen and displayed by:
        I=PLOT(A,B)
and erased by
        I=ERASE(A)
or
        I=UNPLOT(A)
   Communication between the program and the user at the screen is via routines such as
        I=DETAIN (NDET)
This call puts the program into a wait state until an interrupt is received from the 2250. It tells what sort of interrupt was received—for example, light pen, function key, end key, etc. NDET is an array of length 5 that, in the case of a light-pen detect, tells which display variable was detected and the x and y coordinates of the detect. For a function-key detect, the numbers of the function key and keyboard overlay are returned. Routines also exist to use the alphameric keyboard for input and to read such input with the FORTRAN read statement. A more complete description is available from IBM.

III. PERFORMANCE

   Graphic interactive programming encourages a rather unorthodox approach to timing problems. One must consider events in the human time scale as well as in that of the computer. How fast should the computer respond to an attention from the user? Response times of less than 1 sec are hardly distinguishable, whereas times of more than 4–5 sec become uncomfortable, and 15 sec or more almost intolerable. Consider, for example, the rather inefficient method used in the program (dictated by space limitations) for forming a new menu after a detect on a previous menu item. Each item name is read from the control data set and the subroutines to form the order string for the 2250 are called for each. Nevertheless, the response to the detect is virtually instantaneous in human terms.
   Reading cases from the main file and displaying them on the screen is

done at approximately 40 cases/sec, depending somewhat on what is going on elsewhere in the time-shared computer. This is much faster than they can be read by a human, but for a large number of cases can be quite time consuming. The records are not blocked, again because of space restrictions in earlier systems, and a marked increase in speed could be obtained by reading several records at each access. The current direct-access methods available in IBM's FORTRAN IV would require that such buffering be performed in the problem program.

Input/output is the major consumer of time in the program. The calculations of life tables and survival curves are so rapid that the displays appear "instantly" in response to the light-pen detect. Again, though the program allows the user to page back and forth between survival tables and survival curves, space is not used to store these results; they are computed anew for each request.

As noted above, the present main file contains about 10,000 records. It could easily be expanded to contain (for example) the entire California Tumor Registry, which has over 250,000 cases. This would fit easily on one IBM 2316 Disc Pack. However, for a file this large, I/O buffering would become imperative.

IV. CRITIQUE

The approach exemplified by this program is applicable to a wide range of retrieval problems, but not all by any means. As mentioned above, the data to be searched must be in fixed format and numerically coded. For efficiency the file should be used more often for inquiry than for updating. A hierarchical structure with mutually exclusive subcategories renders the logical structure relatively simple.

There are several disadvantages. The fixed format is inflexible, and it is difficult to add or delete variables or categories without reconstituting the entire file. The amount of information is necessarily limited. For example, after finding all cases of lung cancer, a common question is "How many are smokers?" This information is not in the file. Even if it had been included, one never knows when someone may implicate, for example, consumption of chicken fat in carcinogenesis. Had this item not been included in the file in its original design, it would be difficult to confirm or refute the findings.

As applied to the Tumor Registry, the program in its present form has three main uses: (1) as a teaching aid for residents and medical students, (2) to provide general overviews of epidemiologic aspects of tumors, and effectiveness of treatment, and (3) to retrieve suitable cases for a thorough conventional study in which hospital charts and all other pertinent data are reviewed. The danger of using the coded data to form opinions without a study of the original charts must be stressed.

V. DISCUSSION AND ANOTHER SOLUTION

The author wrote this program in this form largely because of the availability of the records already punched on cards.

If one were approaching a complex retrieval problem with a fresh slate, what would he have to consider in designing a total system? Some of the factors would be (1) how the material is to be entered into the computer, (2) whether the coding should be by fixed-field numeric code, by a preselected set of index terms, or by the words of the actual text, (3) what type of inquiries will be directed to the file, (4) what provision will be made for evolution of vocabulary and for types of inquiries not originally envisaged, (5) what proportion of time will be spent in nonproductive file maintenance, i.e., other than data input and inquiry, and (6) how rapidly an answer must be received to a request. Some of the author's answers to these questions are embodied in the following opinions, developed in connection with the problem of retrieval of diagnostic radiology reports.

Physicians detest coding and indexing. If their position is such that they can be bullied into doing it (e.g., interns and residents), they do it perfunctorily and with poor grace. Clerks lack the necessary training to index in the way a properly motivated physician would. In diagnostic radiology it is customary for the physician to furnish an abstract or summary of his report. The words of this summary provide an excellent index to the report. Thus if the summary of the diagnostic radiology report were put into the computer in its entirety—coded internally, of course—it could be searched for key words to retrieve pertinent documents. Some rudimentary syntax recognition would be required, e.g., inclusion of synonyms and subsets, detection of phrases of negation, distance between words, etc. (Lamson and Dimsdale, 1966). However, if every record had to be examined, a search might take several hours.

We would like the file to be interrogated from graphics terminals such as IBM 2260's or with an on-line typewriter. The user should not have to wait more than 10 sec to receive a reply to a request for information. Obviously, sequential tape-oriented files are out of the question. In addition, records will be added to the file at a rate of several hundred per day. The transcription of such records is ordinarily quite costly and time consuming, but the presence of time-shared computers makes it reasonable to have the secretaries on-line. Thus, as they are typing and producing the hard copy for inclusion in the patient's chart, the same information is going directly to the computer for inclusion in the files. The reliability of current time-shared computers is, alas, not such that a busy department could commit itself to the assumption that the computer will be working on any given day. Thus, a back-up system is required. One method would be to punch the reports directly on cards when the computer is down. Hard copy could then be obtained by running the cards through an accounting machine, and when the computer is up, the files could be updated in batch mode.

This large updating activity means that files cannot be maintained in any order other than that of accession number. With the advent of large-capacity direct-access devices, the sorting and merging of files is an anachronism. Files and indices should be open-ended, and maintained as list structures or directed trees. For instance, a dictionary file maintained as a tree could be searched more efficiently than a file ordered strictly alphabetically, because the more frequent words could be closer to the root of the tree. Each vocabulary entry need have a pointer only to the last entry in the main file that contained that word. Each word of the record would have a pointer to the last record that also contained that word. The AND function could be accomplished by picking the pointer that points farthest up the file.

The most frequent inquiries of such a file would probably be by name or hospital number, so especially efficient indices must be maintained for this type of search, and provision must be made for aliases.

Such a project is ambitious, but the author feels that the means are now available to make the computer adapt itself to human communication, rather than the reverse.

References

Lamson, B. G., and B. Dimsdale. (1966). "A Natural Language Information Retrieval System." *Proc. IEEE* **54**, (12).
World Health Organization. (1955). "International Statistical Classification of Diseases," 7th Revision.

CHAPTER 4

Cell Identification and Sorting

L. A. KAMENTSKY[1]

IBM WATSON LABORATORY AT COLUMBIA UNIVERSITY, NEW YORK, NEW YORK

I.	Introduction	107
II.	Characterization of Cells	109
	A. Cell Transport	109
	B. Measurement Characteristics	112
III.	Analysis of Measurements	117
	A. Mass Constituent Representation	118
	B. Classification Techniques	123
IV.	The Rapid Cell Spectrophotometer (RCS)	127
V.	Clinical Applications of the RCS	136
	A. Cancer Screening	136
	B. Viability Assay	141
	C. Differential Blood-Cell Count	142
VI.	Summary	143
	References	143

I. INTRODUCTION

One of the major concerns of clinical laboratories is the identification and counting of biological cells in samples of body fluids presented to the laboratories. The need to automate these tests has been recognized for the last thirty years as their increasing diagnostic value has resulted in wide general use. Simpler tests, such as the counting of erythrocytes or leukocytes of hemolyzed blood, have been usefully automated for some time. Automation of more difficult laboratory procedures such as viability assay, differential blood counts, and detection of cancer cells in exfoliated material or blood is likely to become available to the clinical laboratory during the next few years. A review of methods to implement automation of these procedures is the subject of this chapter.

[1]Present address: Bio/Physics Systems, Inc., Katonah, New York.

Classically, cytological examination involves spreading the cells of interest on a slide and staining and examining them for their morphological distinctions. Since normal methods of proven value are based on morphological criteria as a means of cytological classification, many attempts to automate cytology have been based on mimicking manual procedures by means of a machine. Although our early work was also based on cell-shape recognition, we soon learned that a machine can provide modalities of measurement not practical to the cytologist. I believe the methods of automatically identifying and sorting cells that will come into use will have little resemblance to the classical morphological methodologies.

In this chapter the identification of cells by machine is considered as a pattern-recognition problem. Pattern recognition is the assignment of a meaningful classification to a recognizable structure in a set of signals. A variety of optical or electrical cell properties are available for machine measurement. I will consider only signals resulting from the transduced optical image of the cells. The output of the pattern-recognition machine for each sample will be one of a set of machine codes representing the various classifications predetermined by the machine designer. These classification codes must in turn be presented to the clinician in a meaningful and useful manner.

The total information content of any classification code must certainly be less than that of the transduced image of the cell. It will not reflect most of the detailed characteristics of the cell, nor its position, nor its orientation in the transducer. The pattern-recognition machine thus performs information-destructive transformations on the cell image. Internally, the machine may perform a sequence of transformations to provide new multidimensional representations of the original pattern. Such representations may have a physical meaning. For example, in useful cell identification the nuclear-cytoplasmic-size ratio, the number of nuclear lobes, the presence of cytoplasmic granules, the degree of chromatin clumping, and the amounts of DNA, RNA, and proteins may be informative parameters for generating the classification code.

In Section II, I will describe some methods of obtaining signal representations of cells. Section III is devoted to a description of methods of deriving useful internal representations of the transduced cell images and classification techniques using these representations. The remainder of the chapter describes an application of these ideas to an instrument developed for identifying and sorting biological cells. This device will be referred to as the rapid cell spectrophotometer or RCS. The RCS will be described in Section IV, while Section V will present our experience in using it to automate three tasks: cancer detection, viability assay, and differential blood-cell counting.

II. CHARACTERIZATION OF CELLS

A major advantage of automatic cytological instrumentation over manual screening procedures is the ability to examine every cell of a preparation in turn and apply a defined set of rules for analysis to each cell. Some means must be included in the instrument for isolating each cell or small group of cells and focusing energy on it alone without manual intervention. None of the methods to be described can tolerate averaging of characteristics that would result from characterizing many cells at a time. Three different means of transporting cells with respect to the transducer will be described next.

A. Cell Transport

The following considerations must determine the design of a cell transport system: optical measurement devices have a limited depth of focus, which is a function of the system resolution and capability of automatic focusing devices, if used; the cell-transport system must maintain the cell within the required depth of field of the optical system. The cells must be transported at a rate consistent with the time resolution of the measurement system. The treatment required to prepare cells and the medium in which they are finally examined must not adversely affect their characterization. Differences in light transmission are usually measured between a cell and background. The transport mechanism or medium surrounding the cells must be designed to minimize fluctuations in light transmission not caused by the cells. Finally, consideration must be given to the ability of the transport and preparation system to isolate cells so as to reduce the complexity of the subsequent data analysis.

1. MICROSCOPE SLIDES. The microscope slide has been used universally for transporting cells for visual examination where the cytologist can manually center the microscope stage on each cell of interest. Although there has been successful application of slide techniques for accurate optical cytological measurements with single-cell scanning devices using manual centering (Caspersson et al., 1957), there is no automatic device that uses slides in clinical use as of the time of this writing.

The principal problem in developing an automatic slide scanner is that there is no inherent way in which cells can be isolated from each other. In addition, the boundaries of the microscopic field may intersect the cells, and cell-image reconstruction over multiple fields must be considered. Unless great care is taken in slide preparation there is often a tendency for cells to clump on a slide, increasing the isolation problem. Cell-manipulation devices have been employed in one attempt to develop a slide scanner for cancer screening (Pruitt *et al.*, 1959), but the feasibility of such devices for clinical use has not been shown as yet. Automatic centering and isolation devices have been used in scanners for print reading (Potter, 1964) and may reduce these problems, but they have not been applied to cell identification.

In normal manual observation one is continually changing focus to account for variations in cell thickness, placement of the nucleus, and slide thickness. With good mechanical design and selection of slides it has been claimed by Bostrom *et al.*, (1959) that all cells can be maintained within the depth of field of a high-resolution optical system without manual intervention or automatic focusing.

The fixation of cells to slides and their staining may introduce artifacts, cause variations in staining in different portions of the slide, and introduce variations in background staining that are not present in a liquid transport system. However, automatic staining devices are available for slide preparation and offer a necessary means of quality control in complex staining procedures. The cells on a slide may be maintained in viscous solutions that provide a good refractive match to the cells; the consequent reduction in light scattering may not be possible with a liquid transport system.

The slide is easy to handle and store, although there may be some problem in using it in automatic information-retrieval systems. It is not necessary to make an independent preparation for a machine if slides are used for both manual and automatic cytologic diagnosis. Slides can be marked, or the coordinates of abnormal cells can be automatically recorded in a machine. However, the slide cannot be used in preparative devices for optically sorting cells.

2. LIQUID TRANSPORT. Systems in which the cells are transported in liquid suspension through a narrow channel or aperture have been considered since 1934 (Moldovan, 1934) for either optical or resistivity measurements. Two major advantages have led to widespread use of liquid-suspension instruments for present clinical cell counting: (1) cells may be used with little preparation as they are usually obtained in liquid suspension, and (2) most cells of a sample can be measured in isolation at reasonably spaced intervals without a complex optical–mechanical device.

The flow channel can consist of a capillary tube (Cornwall and Davison, 1960), a channel in a slide (Kamentsky *et al.*, 1965), a metal plate (Kamentsky and Melamed, 1967b) or as an aperture between two chambers (Coulter, 1956). A variety of pumping systems have been employed using vacuum (Coulter, 1956), pressure (Kamentsky and Melamed, 1967b), perastaltic (Technicon Inst. Co. 1966), or infusion pumps (Kamentsky *et al.*, 1965). The cells can be contained in a field as a small as 25 μ^2 if samples are filtered and flow lines are cleaned between samples. Although no high-resolution liquid-flow system has been reported, high-resolution measurements can be made by causing the cells to flow in the direction of the optical axis so that each cell passes through the focal plane.

In addition to the advantage of easy isolation of cells, a major advantage of cell flow is that the optical system is fixed and all imaging is at the principal axis of the lenses so that there is no need for a separate reference system. Background measurements to compensate for slow lamp and power-supply fluctuations are made between cell events with identical optical conditions for the cell and for the background. As will be seen later, simple low-drift ac-coupled photodetector amplifiers can be used to directly generate the difference between transmission with and without a cell.

Different techniques for cell fixation and staining are required for dealing with suspended cells. Strengths of solutions can be changed gradually or reagents added sequentially. Centrifugation steps are required to wash the cells when using interacting dyes or reagents. Unfixed viable or dead cells can be used directly in liquid suspension with stains just as they are in a counting chamber. This may simplify procedures for certain applications without loss of sample as would occur in using slides without fixation.

A major difficulty with liquid transport is the problem of matching the refractive index of the cells with that of the suspending medium to minimize light scattering. Cellular scatter is thus a part of any optical measurement of absorption and must be considered in the analysis of the measurement data. However, scatter itself may be a useful measure to estimate cell size (Kamentsky et al., 1965).

A second problem of liquid transport is clogging of the flow channel. Care in sample and reagent preparation to prevent extraneous material from entering the flow is essential, and prefiltering is necessary for certain samples. We have found that with reasonable care clogging is a rare event with flow channels as small as 100 μ^2 and more frequent but tolerable with blood or cell-culture samples to 25 μ^2 channel cross sections. However, a rapid and automatic system to flush the channel in reverse for cleaning between samples and when a clog occurs is an essential part of a useful instrument.

Cell sorting has been achieved using a liquid transport and an electrostatic deflection system adapted to the Coulter principle by Fulwyler (1966). A fluid switch has been adapted to an optical measurement system by Kamentsky and Melamed (1967b). An oscillograph in which droplets of ink are deflected by an electrostatic field was developed by Sweet (1963) and adapted by Fulwyler and studied in our laboratory as a means of sorting cells. We found that a fluid switch (described in Section IV) was more adaptable to an optical measurement system.

3. TAPE TRANSPORT. The tape transport combines some of the features of the slide and the liquid transport to provide a continuous stream of cells fixed to a transparent tape. It has recently been developed by Tylko *et al.* and described by Boddington and Diamond (1967) for automatic analysis of exfoliated material. A "graphing machine" places a fine line of cells on a tape. This tape may be passed through staining baths; the cells are covered with a mounting medium and subsequently covered with Scotch tape. The tape can be marked to facilitate automatic sample identification and to mark off cells for later automatic access to abnormal cells. Tape length for each sample is about one meter. It is too early to evaluate the quality of the preparation, the extent of cell isolation, or background variations encountered in the method.

B. Measurement Characteristics

1. HIGH-RESOLUTION MEASUREMENT. An optical scanner is required to convert an image of a cell into a sequence of electrical signals. Many different cell scanners have been described. A good review of some of these can be found in "Scanning Techniques in Biology and Medicine" (Montgomery, 1962). In each case the cellular image is represented by a set of values describing some optical properties at discrete points on the cell image. To achieve cell identification, these values may be converted into digital form and stored on magnetic tape or in a computer. I believe that studies involving high-resolution scanning and multilevel quantization of scanned data may yield fundamental cytological information and are important to many research uses. However, the direct application of this technique to clinical devices where large numbers of cells must be processed in a few minutes presents serious problems.

The signal-to-noise ratio of any scanning device determines the number of meaningful levels or nonredundant bits that can represent each image element. The signal-to-noise ratio is physically limited to $\sqrt{N}$, where N is the number of photons per measurement. N is directly proportional to the product of the intensity of the scanning spot and the time that that spot dwells on each image element. We have found that practical scanning systems may be severely limited in signal-to-noise ratio by light limitation at scanning rates above 10^6 bits per second (bps).

Of greater concern is the capability of the data-analysis unit that must process the scan data. Each data bit may have to be manipulated many times to carry out pattern-recognition algorithms. Special-purpose parallel-processing arrays have been devised for pattern recognition (Unger, 1958;

Kamentsky, 1959). Unless such special-purpose integrated-circuit arrays are developed for morphological pattern analysis, processing rates higher than 10^6 bps will be too costly for clinical use.

A practical device may be required to analyze 100 to 1000 cells each second so that samples of 10,000 to 1 million cells can be processed in a few minutes. A successful recognition technique that has been studied with data from scan rates of the order of 10^3 or 10^4 bps may be found to be physically unrealizable or unusable in processing the low signal-to-noise ratio data obtained at the scan rates of 10^6 to 10^8 bps. Such high bit rates will be found if the scanner resolution is near the limit of optical resolution and cell rates of 10^2 to 10^3 per second are required.

At least three possibilities exist for modifying the scanner to reduce the data rate. One possibility is to modify the scanner to generate binary data that can adequately represent cellular morphology so that multilevel representations are not required. Figure 4-1a is a photograph of a cell field. This field was scanned with a cathode-ray-tube scanner in our laboratory. The scanner signals were quantized into two levels to obtain binary representation of the cells. Figure 4-1b is a photograph of the scanner display of Fig. 4-1a. Using a technique described by Gelerenter *et al.* (1966), the focus electrode of the cathode-ray tube was modulated to vary the scanning-spot size. The in-focus and out-of-focus signals were subtracted, and the difference signal was quantized into two levels again. The result of scanning Fig. 4-1a in this manner is shown in Fig. 4-1c, which shows considerably more morphological characteristics of the cells than does Fig. 4-1b.

Another method of data reduction is coming into wider use with the availability of programmed scanners in which the scan pattern is computer controlled. The programmed scanner can modify the position, resolution, or quantization level of the scan as a function of the information previously scanned. The image itself is used as the storage medium, so that sequential decision procedures may be implemented without scanning the complete cell image. This technique has not been applied to cell identification as yet, although it has been used by Kamentsky (1961) for recognition of handwritten numbers and by Potter (1964) for print recognition. The philosophy of use and implementation of a two-wavelength programmed scanner is more completely described by Highleyman and Kamentsky (1959).

A third alternative to data rate reduction is to decrease the scanner resolution. The use of a single-spot or zero-resolution scan is developed next.

2. ZERO-RESOLUTION MEASUREMENTS. The scanning spot may be increased in area until it is larger than the size of any cell. Cells intersecting

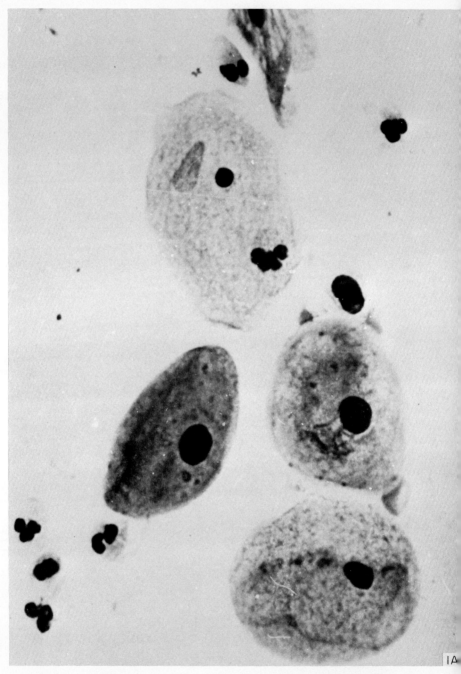

FIG. 4-1. Display photographs of binary images of a field of cells (a), quantized into two levels based on optical density (b), and the gradient of optical density (c).

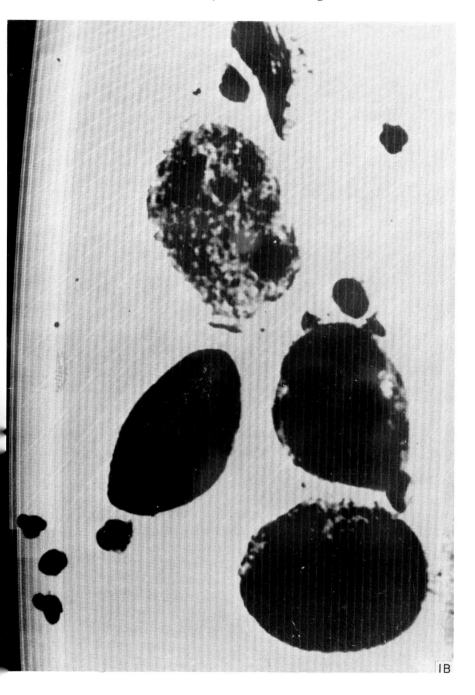

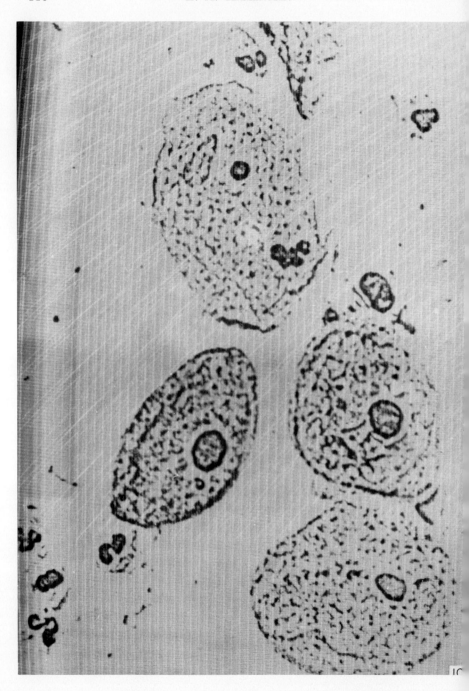

4. Cell Identification and Sorting

the scan spot will diminish the total transmission of the incident energy as a function of their absorption and scattering cross sections. Alternatively, fluorescent or dark-field optical systems can be used whereby there will be an increase in light in the optical system as the scanning beam and cell intersect. Cells have absorptions at specific wavelengths, which can be related to cellular constituents. They can also be stained with a great variety of dyes to produce absorption or fluorescent signals proportional to many different cellular constituents. Thus, with a multiwavelength optical system, as will be described in Section IV, several different cellular constituents can be simultaneously estimated.

The depth of field of an imaging system is a function of the required resolution. Collimated light is used in the zero-resolution system, so there is no requirement on position of the cells along the scanning beam. This greatly simplifies the cell-transport system requirements.

Since the scanning spot is as large as the cell, the energy available for a measurement may be increased by a factor of the order of 10^3 over higher-resolution scans. The measurement rate can be correspondingly greatly increased and still yield sufficient signal-to-noise ratios. We have obtained signal-to-noise ratios as high as 100 to 1 with cell rates in excess of 1000 cells per second. Thus each cell is characterized by about 7 bits per wavelength measurement. With the four-wavelength measurement system presently in use the information rate of 3×10^4 bps can be conveniently processed by special-purpose circuits or a small-scale computer.

III. ANALYSIS OF MEASUREMENTS

With the exception of computer-controlled scanning, the characterization methods described in the previous section will each yield a set of simultaneous data values. If there are p such data points each quantized into q states, any data-analysis system must assign each of q^p possible data patterns to a specific classification. It is clearly impractical to estimate and store the information required to directly assign all states of the characterizer. Because of this the data are analyzed using a sequence of logical manipulations to develop a relatively small number of new parameters. This new data pattern should be stable to changes in the position, orientation, size, or shape of the cell that are *not* relevant to its identity. These parameters can be considered to form the coordinates of a space. Many different and simply implementable geometric constructs are available for partitioning volumes in such a space. The recognition-logic designer must find a parameter space in which almost all of the set of parameter values belonging to each class will overlap clusters of other classes as little as possible. To my knowledge there is only one example where these principles have been applied to computer cell recognition. Prewitt and Mendelsohn (1965) developed a program to distinguish four classes of leukocytes using a 40,000-point, 8-bits-per-point representation of 50 sample cells.

We attempted to apply these principles to patterns developed by a modulated-spot scanner described in Section II. We used a computer-automated technique for designing recognition parameters developed for multifont print recognition (Kamentsky and Liu, 1963). On the basis of experience with the development of practical automatic handwritten-numeral and multifont-print readers, I believe that recognition procedures could be developed to distinguish a small fixed set of design images. However, this logic works poorly on images not used in the design of the recognition logic. The design of systems to reliably classify large samples of new images involves processing thousands of design samples, with concurrent large manpower and computer expenditures. In the case of cell identification, ten times as many data are required to represent each cell as are needed for print recognition. The required manpower and computer time to develop a cell-recognition logic by computer simulation are too great, and we sought other characterization techniques requiring many fewer data.

We strongly felt that the use of zero-resolution measurements would more quickly result in a practical clinically usable cell identification system. However, the data obtained from zero-resolution measurements are known not to be independent because of overlapping absorptions and light scattering of multiple constituents. In this case there is a systematic method of data reduction that can yield estimates of constituent masses as descriptive parameters. This is described next and is followed by a discussion of some geometric constructs we are using for final classification of results.

A. Mass Constituent Representation

Both spectrophotometric and chemical quantitative analysis have been used extensively to characterize differences in the mass of certain constituents in biological cells. Data so obtained may be used directly to establish a basis of cell identification. The cytochemical literature may be utilized as the basis for chemical treatment and spectrophotometric techniques to determine any of a number of cellular constituents. I will show in this section that a zero-resolution multiple-wavelength spectrophotometric scanner can give estimates of mass constituents and, as will be seen, can also provide estimates of the nuclear area in certain cases. In the second part of this section I will show that fluorescence measurements provide direct indications of mass constituents in a zero-resolution spectrophotometer.

4. Cell Identification and Sorting

1. ABSORPTION MEASUREMENTS. Consider a cell spectrophotometer in which the object plane is illuminated over an aperture area A with uniform parallel light having a broad spectral range. With no cell in the aperture a total radiant flux of $U_{0\lambda}$ in a band about the wavelength λ can be collected by each of n photodetectors sensitive to wavelength λ. The aperture A is large enough so that when a cell is transported into the aperture, the cell is completely contained within the volume defined by A and some fixed length along the optical path Z. The cell constituents or dyes bound to these constituents will absorb or deflect the incident energy to reduce the total flux at each photodetector to U_λ. A cell spectrophotometer is considered in which n photodetector-signal amplifier outputs S_λ give the difference between the photodetector signal when a cell is completely contained in A and background. Thus

$$S_\lambda = G_\lambda(U_{0\lambda} - U_\lambda) \qquad \lambda = \lambda_1, \lambda_2, \ldots, \lambda_n \tag{1}$$

and G_λ is the ratio of signal voltage to radiant energy for each photodetector and associated amplifier.

The total effective radiant energy collected by photodetector λ is the integral of the flux density $I_\lambda(x,y)$ at all points x,y of an image plane of aperture A, since the flow-channel illumination is parallel and I consider both the reference and signal flux with respect to the same plane. Thus

$$S_\lambda = G_\lambda \iint_A \left(\frac{U_{0\lambda}}{A} - I_\lambda(x,y) \right) dx\, dy \tag{2}$$

if $I_\lambda(x,y)$ is due only to a cell completely contained in aperture A. The extinction $\tau_\lambda(x,y)$ at each point in an image plane resulting from light loss by the cell is

$$\tau_\lambda(x,y) = \log_{10} \frac{I_\lambda(x,y)}{U_{0\lambda}/A} \tag{3}$$

By Lambert's law the extinction of light in the z direction $\tau_\lambda(x,y)$ at the point x,y is the integral along the optical axis of the extinction $\tau_\lambda(x,y,z)$ due to each volume increment $dx\,dy\,dz$ at position x,y,z or

$$\tau_\lambda(x,y) = \int_0^Z \tau_\lambda(x,y,z)\, dz \tag{4}$$

Thus at each point x, y the photometric field of aperture A can be considered to have an extinction $\tau_\lambda(x,y)$ for light in the z direction. The integral of this quantity taken over the aperture A defined as τ_λ will be shown to be directly proportional to a function of each of the constituent masses M_i of the cell and its nonspecific losses. Since the signal obtained in each of the photodetector amplifiers S_λ is not in general proportional to τ_λ, there are two problems: (1) to relate τ_λ to the constituent masses M_i, and (2) to relate the total absorption to τ_λ. I consider first the problem of relating the extinctions to the set of masses.

If the concentrations of light absorbing or deflecting molecules is in the range where Beer's law is obeyed, the extinction due to any constant concentration incremental volume $dx\,dy\,dz$ is linearly related to the concentrations of absorbing molecules in the volume. If the probability of scattering light back into the detectors is small once it has been deflected out of the detectors' numerical aperture, then the extinctions observed by the detectors due to an increment of volume $dx\,dy\,dz$ is

$$\tau_\lambda(x,y,z) = \sum_{i=1}^{m} K_{\lambda i} C_i(x,y,z) + \Phi_\lambda(x,y,z) \tag{5}$$

where there are m constituents of concentrations $C_i(x,y,z)$ at point x,y,z. $K_{\lambda i}$ is an extinction coefficient for constituent i and wavelength λ, and $\Phi_\lambda(x,y,z)$ are the nonspecific light losses at wavelength λ.

Combining Eqs. (4) and (5) and integrating with respect to x, y, and z,

$$\sum_{i=1}^{m} K_{\lambda i} M_i = \tau_\lambda - \Phi_\lambda \tag{6}$$

since the mass M_i of constituent i is the integral of the concentration of i over the observation volume and I define

$$\tau_\lambda \equiv \iint_A \tau_\lambda(x,y)\,dx\,dy \tag{7}$$

$$\Phi_\lambda \equiv \int_0^Z \iint_A \Phi_\lambda(x,y,z)\,dx\,dy\,dz \tag{8}$$

The nonspecific light losses Φ_λ are a function of the state of cell constituents and the position of these constituents with respect to the cell-spectrophotometer optics and must in general be reduced to calculate constituent masses. The nonspecific light losses are usually reduced by choosing a suspending medium whose refractive index is close to that of the major cellular constituents. However, this is not always possible: (1) there may in certain cases be an anomalous dispersion due to major constituents and their interfaces that cannot be matched by any suspending medium that will not absorb light at the measuring wavelength, and (2) the required suspending medium may not be compatible with the cell-transport system or may affect the binding or color of the dyes used to stain cells. Thus it is useful to develop a technique for calculating the contribution of nonspecific losses.

The basic assumption applied to calculate the contribution of nonspecific losses is that although their dependence on wavelength may be complex, the ratio of nonspecific light losses at two given wavelengths is the same constant for all cells. The cells are always fixed alike and suspended in the same medium. A dark-field optical system in combination with a bright field may provide a more accurate determination of scatter and can be used alternatively. This assumption is illustrated and shown valid for a tumor-cell suspension by Kamentsky and Melamed (1967a). Thus

$$\Phi_\lambda = a_\lambda \Phi \tag{9}$$

where a_λ is a constant depending only on wavelength and the optical system, and Φ is a constant for a specific cell.

Equation (6) then becomes

$$\sum_{i=1}^{m} K_{\lambda i} M_i + a_\lambda \Phi = \tau_\lambda \qquad \lambda = \lambda_1, \lambda_2, \ldots, \lambda_n \tag{10}$$

If a_λ is known, $m+1$ measurements of τ_λ will determine the m mass constituents and a relative value for Φ with $a_n = 1$. The constants a_λ are determined initially by performing an experiment so that the term $a_\lambda \Phi$ is large compared to all $K_{\lambda i} M_i$. The cells are measured undyed if visible dyes are to be used for constituent determination, or cells low in nucleic acid or with the nucleic acid extracted are measured in the ultraviolet at the same wavelengths to be used in rapid determination. Then

$$a_{\lambda i} = \frac{\tau_{\lambda i}}{\tau_{\lambda n}} \bigg|_{K_{\lambda i} M_i / \Phi \to 0} \tag{11}$$

I will next describe two methods we are planning to use to relate the total absorption measurements yielding the set of signals S_λ to the mass constituents. In the first the ability of an instrument to work at low extinctions with high incident energy is exploited. The method of background referencing in a flow system allows determination of low extinction values that may not be possible with other methods. Cells may be understained or measured away from their absorption maxima to achieve low extinctions. Combining Eqs. (2) and (3),

$$S_\lambda = \frac{G_\lambda U_{0\lambda}}{A} \iint_A [1 - 10^{-\tau_\lambda(x,y)}] \, dx \, dy \tag{12}$$

If $\tau_\nu(x, y) \ll 1$,

$$\tau_\lambda = \frac{S_\lambda A}{2.3 \, G_\lambda U_{0\lambda}} \tag{13}$$

Thus the signal voltages S_λ from each of the photomultiplier amplifiers may be used directly for τ_λ in Eq. (10) and $m+1$ wavelengths will determine m constituents if wavelengths are chosen so that the extinction of the cell or dyes bound to the cell is small and the signal-to-noise ratio is sufficient. The validity of this assumption can be checked by a second set of wavelength measurements.

A second approach to relating τ_λ to S_λ is to postulate a specific distribution for cell constituents first introduced by Ornstein (1952) and by Pateau (1952) and developed by Mendelsohn (1958). Equation (12) can then be integrated and the ratios of two S_λ used to determine mass ratios as shown by Kamentsky and Melamed (1967a), or two wavelengths can be used to solve for a single value of τ, and a mass density.

In many interesting cell models the distribution of scattering molecules is different from the distribution of absorbing constituents. This necessarily will complicate any calculation of constituents using a technique where specific distributions are postulated if nonspecific losses are appreciable. A model has been carried through using different areas for scatter and absorption and has led to complex solutions. Thus the following methods are only applicable where there is a proper refractive match. A second restriction, that there be no overlap of absorption spectra at the measuring wavelengths, is required to simplify this presentation.

In Eq. (12) let $\tau_\lambda(x, y) = K_{i\lambda} M_i / F_i$ in area F_i and let $\tau_\lambda(x, y) = 0$ elsewhere in aperture A. Then, if all of the masses M_i are confined to projected areas F_i,

$$S_\lambda = \frac{G_\lambda U_{0\lambda}}{A} F_i (1 - 10^{-K_{\lambda i} M_i / F_i}) \tag{14}$$

The two-wavelength method of Ornstein and Pateau can be applied directly to Eq. (14) to solve for the values of M_i and F_i for each constituent using two different wavelength measurements of S_λ. As Ornstein and Pateau have done, the two wavelengths may be selected so that $K_{\lambda_1, i} = 2K_{\lambda_2, i}$ directly, or solution tables can be generated for any ratio. It is important to note that the projected areas F_i can provide useful morphologic information for cell identification. In certain studies the approximate mass of a constituent may be known (such as the DNA of normal leukocytes). Under these conditions the projected area of this constituent F_i can be determined from one measurement and may provide a useful identification parameter in combination with other measurements.

4. Cell Identification and Sorting

2. FLUORESCENCE MEASUREMENTS. The distributional error and nonspecific light losses can be made negligible by using a cell spectrophotometer to measure fluorescence of dyes bound to constituents. If the wavelength used to excite fluorescence does not overlap the wavelength range used to measure emission, there will be no contribution to the measured emission by nonspecific scatter. Alternatively, the scatter emission has been measured independently using dark-field illumination to provide additional morphological information.

The elimination of distributional error requires negligible absorption of the incident energy and fluorescent emission by cell constituents or dyes. If fluorescence is excited where there is low extinction and if emission is measured at a wavelength where there is again negligible extinction, the signal voltage from a cell spectrophotometer for cells in a fixed optical position is

$$S_\lambda = G_\lambda \sum_{\lambda'=\lambda'_1}^{\lambda'_l} \frac{U_{0\lambda'}}{A} \sum_{i=1}^{m} Q_{\lambda\lambda',i} M_i \qquad (15)$$

where λ' are the excitation wavelengths and λ is the emission wavelength, and $Q_{\lambda\lambda',i}$ is a coefficient relating the intensity of fluorescence at wavelength λ, with excitation at the wavelength λ' and the dye bound to mass constituent i. Thus the spectrophotometer signal voltages S_λ are directly proportional to the cell mass constituents M_i.

If cells are measured in a flow channel, the amount of energy collected will be dependent on the position of the cell in the aperture volume because of obscuring of radiated energy by the channel walls. The fluorescent signal is very position dependent, and transport systems to measure fluorescence must be designed to minimize the variation in the collection efficiency of the optical system.

B. Classification Techniques

The final step in cell identification is to assign each point in the m-dimensional space representing the m estimated mass values to a predetermined category. The space is partitioned by some algorithm that must contain a set of parameters estimated by studies of properly identified data. A simple example of this is illustrated in Fig. 4-2, which is a two-dimensional histogram from our spectrophotometer of a vaginal wash from a patient with cervical cancer. In this printout the vertical coordinate represents total cellular absorption measured at 2600 Å, while the horizontal coordinate represents cell scatter at 4100 Å. The line shown in the figure has been chosen using samples of normal epitheleal cells and cervical

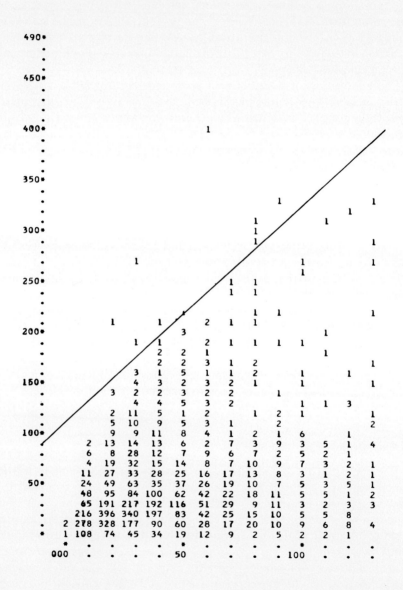

Fig. 4-2. Two-dimensional histogram printout from RCS-1130 of cells obtained coordinate is absorption at 2537 Å; the

4. *Cell Identification and Sorting* 125

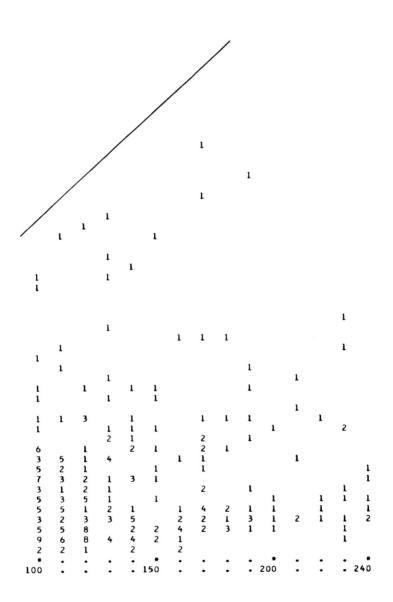

by vaginal wash of a patient with carcinoma of the uterine cervix. The vertical horizontal coordinate is scatter at 4000 Å.

cancer cells so as to best place the cancer cells above the line and the normal cells below it. Thus the parameters needed to describe this line define a classification scheme for partitioning a mixed sample into cells called normal and cells called abnormal. It must be realized that biological cells may form a continuous spectrum of properties. It follows that no classification scheme, including that used by the pathologist, will be perfect, particularly for classifying cancer cells. Parameters that define classification can only be determined on a basis that will cause the least misclassification of cells in a large sample.

I have illustrated an example where a simple curve is used to partition a two-parameter measurement space. The use of higher-dimensional surfaces to partition higher-dimensional measurements has perhaps been the most highly developed method of classification. Many classification schemes can be reduced to hyperplane partitioning because of the smaller number of parameters needed to describe such hyperplanes. In any practical implementation of a classifier, the designer must consider the problems of estimating, storing, and operating with the separating parameters that lead him to simpler descriptions. More complex separation boundaries can sometimes be considered in terms of sets of hyperplanes.

Another useful classification that can be illustrated simply is the concept of distance functions. A point in a multidimensional space may be assigned in terms of its distance from each of one or more points or surfaces in that space. Figure 4-3 illustrates this and and is an example of an unprocessed display of peripheral blood cells using stains specific for DNA and certain proteins. Both a dot-pattern display (Fig. 4-3a) to be described later and a computer printout (Fig. 4-3b) are shown. The vertical coordinate represents absorption at the maximum of the protein stain and the horizontal coordinate absorption at the DNA stain maximum. In this case there is a set of discrete clusters where each dot represents a cell. We know that the major clusters represent erythrocytes, lymphocytes, and granulocytes. To classify these data we may establish one or more representative points on the display for each cell class. On an unknown sample the coordinates of each cell are measured and the distances of its point in the space computed with respect to a predetermined point or surface representing each class. It is assigned to the class having the minimum distance.

Figure 4-3 also can be used to illustrate still another classification scheme. The measurement space can be partitioned in terms of hypervolumes. In the simple case illustrated, closed curves can be defined and cells classified as a definite category if they fall within the areas defined by these curves. Methods have been developed using volumes that can be defined by a reasonable set of parameters such as hyperspheres.

These three classification schemes illustrate a small part of a considerable literature on this subject. Sebestyen (1962) and Highleyman (1962) have written general treatments of classification processes and their implementation.

There is of course no assurance that data are separable in terms of a given set of descriptors or that any simple classification scheme will establish separability. There has been much work using another point of view, that of developing methods of automatically establishing parametric distributions that will cluster data (Bonner, 1965), or make the parameters more informative and independent in terms of information or other measures (Kamentsky and Liu, 1963). Such transformations of data can be made to yield displays of lower dimensionality. We are presently studying spectrophotometric data by automatically developing transformations that will cluster these data in a lower dimensionality. It is of interest to determine whether these transformations will be similar to those of the mass-constituent analysis.

The resulting classification can be used to differentially count cell types or actuate a cell-sorting or cell-marking mechanism. Our experience has been that for certain problems it may be more useful to the clinician to provide a display of unclassified parametric data, using a two-dimensional histogram or contour map as will be described in the next section. Spectrophotometer measurements not only may provide systematic analyses of many cells but they may provide entirely new methods of assaying clinical material. We do not want to discount the use of spectrophotometrically determined display patterns as a useful record to be used for screening or diagnosis. In certain cases such as chromosome karyotyping it may even be advantageous and more economic to develop an interactive system between a scanner–computer and an operator so that the more difficult pattern-recognition procedures are handled manually.

IV. THE RAPID CELL SPECTROPHOTOMETER (RCS)

An instrument is described in this section for measuring and recording zero-resolution optical properties of biological cells flowing in liquid suspension at rates up to 1000 cells per second. This instrument was designed to characterize each cell in populations of the order of 100,000 cells or more in a few minutes, so that even a few cells with desired characteristics could be identified. Cells may also be physically separated as a function of their optical properties by a fluid switch in the flow channel.

The rapid cell spectrophotometer shown diagrammatically in Fig. 4-4 consists of a microscope with a flow channel mounted on the stage and a series of photomultipliers with associated amplifiers and circuits to

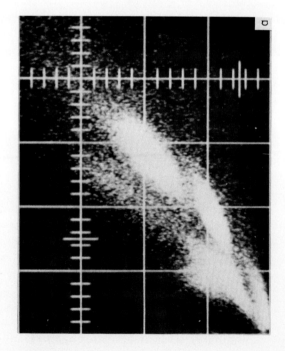

Fig. 4-3. Display photograph (a) and two-dimensional histogram printout (b) from RCS-1130 of human peripheral blood cells stained with Feulgen and Naphthol Yellow S. The vertical coordinate is absorption at 4300 Å; the horizontal coordinate is absorption at 5700 Å.

4. *Cell Identification and Sorting* 129

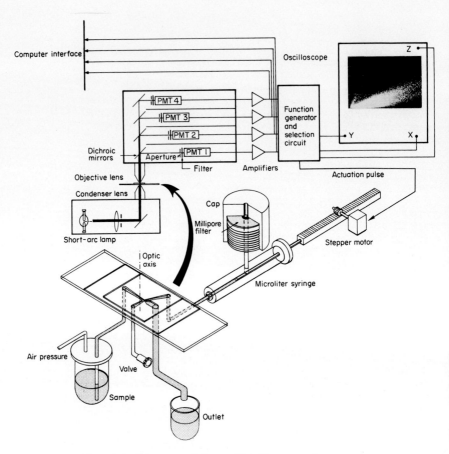

Fig. 4-4. Diagram of the rapid cell spectrophotometer.

develop signals proportional to the absorption, scattering, or fluorescence of cells flowing through the channel. The signals are displayed as a pattern of dots on an oscilloscope, or are analyzed by an IBM 1130 computer through an interface. The signals can also be processed to activate a cell sorter.

The cells in suspension are pumped through the channel at flow rates up to 0.5 ml/min by pressurizing a plastic sample container. The channel is produced by photoetching the flow pattern into a molybdenum plate that is then cemented between a quartz slide and coverslip. The minimum channel width and plate thickness used have been 25 μ or 100 μ. Polyethelene tubes are cemented into holes cut through the slide at the end of each port of the flow channel illustrated in the figure. During cell sorting, which is described more completely by Kamentsky and Melamed (1967b), the cells are observed at the optical axis by the spectrophotometer and flow to the sorter junction in 2 msec. Unselected cells continue to flow out into the outlet container. If a specific cell has optical properties causing an actuation signal, the stepper motor is pulsed, causing the plunger in the syringe to draw 0.03 μl. of fluid. This fluid pulse is propagated across the flow channel at the junction, causing fluid flow across the junction for about 3 msec. There is a delay of 2 msec from the time of detection until an appreciable flow occurs across the channel, matching the delay time of the cell to reach the junction from the observation point.

In practice the cells flow at varying velocities, owing to the parabolic velocity profile across the channel. The velocity profile of the cells can be measured by imaging the cells upon a grating, and the actuation delay and displacement time can be matched to this profile. The actuation time can be made as long as necessary to ensure that the desired cell has been captured in the side channel. In this case, several other cells will accompany the selected cell. We were successful in achieving a final concentration of selected cells of about 1:5 from initial concentrations in the range of 1:10,000. The selected cells are drawn far enough into the side channel to remain there for the duration of a run. Up to 300 selections can be made during a run of 100,000 or more cells.

After a sample has been processed, the main channel is automatically flushed in reverse for cleaning. A Millipore filter is placed on a holder connected by tubing to a radial hole drilled into the barrel of the syringe. During the wash the plunger of the syringe is withdrawn to just beyond the radial hole so that there is a flow of clean water through the side channel, the syringe, and the Millipore filter. The separated cells are thus flushed out of the side channel and trapped on the Millipore filter. The Millipore filter is then removed, placed on a slide, and stained for visual observation of the cells. The cells are all visible in one low-powered microscope field, as illustrated in Fig. 4-5.

The spectrophotometer light source for studies of stained cells absorbing visible light is a xenon short-arc lamp or a mercury lamp for ultraviolet or fluorescence studies. Using Kohler illumination, the source is imaged onto the back plane of a condenser lens at a low numerical aperture such as to provide nearly parallel illumination through the channel. Apertures both at the source lens and at the objective-lens image plane in front of the photomultiplier tubes limit the field on the flow channel. Dichroic mirrors reflect successively longer wavelengths of light to successive modular assemblies consisting of an aperture, an interference filter, and a lens imaging the source onto a photomultiplier tube. The light at each photomultiplier will be reduced or increased during 200 μsec as each cell flows

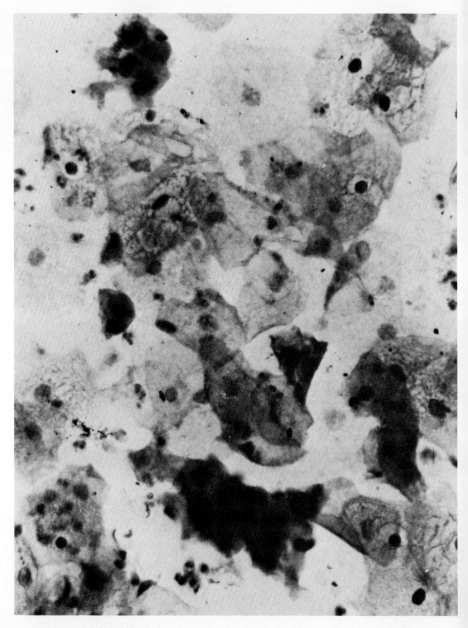

FIG. 4-5. Cells with high ultraviolet absorption selected from a vaginal wash specimen of a patient with carcinoma of the uterine cervix. The Millipore filter was stained with hematoxylin and eosin.

past the optical axis. Bright-field optics can be used to measure absorption and fluorescence. Alternatively, dark-field optics have been used to measure fluorescence and scatter.

The photomultipliers are regulated to a fixed anode current by automatic control of the dynode voltage. The amplifiers are bandlimited to eliminate all photomultiplier noises above 10,000 Hz. To eliminate lamp-supply and arc fluctuations, vibrations, and line noise, the dynode amplifier frequency–gain characteristic is shaped to regulate out all fluctuations below 200 Hz. The regulator circuit does respond to some extent to the cell signals. The effect of this finite low-frequency cutoff is partially corrected by circuitry and may be further reduced by the 1130 computer. The major advantage of feedback control of the dynode voltage is that the background level between cell signals is effectively subtracted out, yielding S_λ of Eq. (1) directly from the photomultiplier amplifiers.

The peak cell signals representing the absorption or fluorescence when a cell is completely in the aperture are detected for each cell by differentiator and zero-crossing circuits. These peak signals transmitted to the 1130 interface, to the cell-selector activation circuit, and to a set of analog operational amplifiers to transform the signals or combine them to produce two signals to deflect the two axes of an oscilloscope. Each cell thus produces a line on the face of the oscilloscope. The end point of this line (which has coordinates given by the magnitudes of the two signals) is intensified by a 1-μsec pulse. The beam intensity of the oscilloscope is set so that only a single dot appears on the screen for each cell. All such dots produced by a sample are photographed to produce a display such as shown in Fig. 4-3.

Two cell-sorting classification functions are built into the RCS. The first selects cells with respect to the line $y = ax + b$, where x and y are the display coordinates and a and b are parameters that can be set on potentiometers. Cells on each side of this line are counted, and the signals above the line can actuate the cell sorter. A second classification criterion built into the RCS is an area selector, in which four potentiometers specify the limits of a rectangular area on the display.

The RCS has been interfaced with an IBM 1130 computer to analyze spectrophotometric data on-line, since use of a computer provides several advantages. The data-analysis techniques described in the previous section can be implemented with multiple measurements. With this system the experimenter can communicate with the data, using the console typewriter to call up programs, to implement and evaluate a variety of different computational techniques. The data may be displayed as a printed two-dimensional histogram as shown in Fig. 4-3 or as a contour map using an IBM 1627 plotter, or lists of properties or counts for selected cells may be printed on the console typewriter. The 1130 magnetic-disk file can be used to store and manipulate data from many samples. Sample identification and bookkeeping functions are handled automatically. The computer has access to any of 16 voltages in the RCS so that gain factors are computed automatically independent of the RCS control-dial settings. Finally, the interface can control up to seven functions in the RCS to initiate cell sorting or other mechanical operations.

The interface utilizes the cycle-steal features of the 1130 to provide overlap of data aquisition, disk storage, and computation. Cell data may be grouped into conveniently sized records for packing and storage on disk. An input/output command causes the interface to read a set number of cell amplitudes into core and then interrupt the CPU. The specific photomultipliers used, the number of cells per record, and the core data address are all programmable. The number of cells per record is set to 1 during a sorting run so that the CPU is interrupted after each cell. The CPU is also interrupted by a change in status of the RCS such as the start of a sample. CPU interruptions cause a transfer of the program to an interrupt service routine, which examines the state of a device service word generated by the interface. This word contains the information necessary to the program to indicate that the RCS needs service and the type of service required.

The specific cell data transmitted to the 1130 are the peak values of the four photomultiplier amplifiers, the pulse width with respect to a fixed threshold of the conjunction of the four signals, and the time between cells. The pulse width provides an additional measure of low rate and cell clumping. The time between cells is useful for correcting for low-frequency distortion.

A separate analog-to-digital converter is used for each photomultiplier in the interface. They produce an 8-bit binary representation of each peak signal amplitude in 80 μsec. Conversion is activated by a signal on any of the four inputs to sample and hold amplifiers. The signal is held at the peak value by these amplifiers until the conversion is completed. The time between signals and the pulse width are both generated by gating a clock to two counters. The counting time and data transfer to a buffer register are controlled by the cell peak signal in one case and the width signal in the other. The time between cells is transmitted to the CPU when a cell is detected. After conversion the four peak values are transmitted in successive cycle-steal operations. The pulse width is the last data item transmitted and is gated to the 1130 when all the cell signals pass below their thresholds.

The interface is provided with a 16-input multiplexer and a fifth analog-to-digital converter. These inputs can be connected to points in the RCS to measure static voltages under program control. A write command sets the multiplexer address and initiates a conversion. The device-status word, which contains a bit indicating the end of conversion, can then be interrogated by the program. The voltage data can be read and stored by the program when conversion is completed, using a command to read the data word to a specific core location.

An identification number generated in the RCS can be read under program control by using a sequence of write commands to identify each RCS identification register position, followed by a read command to read each register status and store it in core. Changes in RCS status due to the beginning or end of samples cause a CPU interrupt and create specific bit configurations in the device-status word. The interface can generate up to seven control actions initiated by an input/output control command. Two of these are used at present: one to initiate a calibration mode and one to pulse the sorter motor.

FIG. 4-6. Photograph of the RCS, IBM 1130 computer, and computer interface.

Figure 4-6 is a photograph of the RCS, the interface unit, and the 1130 console. A major portion of the RCS shown here provides for automatic sample handling and sample identification. Samples in plastic cups are automatically stirred and fed sequentially into the cell spectrophotometer using a 30-sample turntable. The flow channel and all parts in contact with a sample are washed and air dried between sample runs. Circuits are incorporated in the RCS to detect a clog and wash the channel without advancing the sample and to detect dense samples and reject the sample in this case. Sample runs are ended when the container is empty, a set number of cells is counted, or a set time limit is exceeded. The RCS can control a keypunch directly to produce punched cards.

V. CLINICAL APPLICATIONS OF THE RCS

The RCS has been used in preliminary studies to develop a variety of clinical and research applications of this technique for cell identification and sorting. In our laboratory we have studied (1) detection of cancer cells in exfoliated material or blood, (2) differential counting of blood cells, (3) viability assays, and (4) cell differentiation. These studies are in different stages of their proof of feasibility. None can be considered, as of this writing, as practical enough for routine clinical laboratory use. I will describe the methodology, present performance, and future plans for each of the three clinical laboratory applications.

A. Cancer Screening

The RCS was conceived as an instrument to screen samples of exfoliated material for detection of cervical cancer. It was thought valuable to develop a device that could select a reasonably small fraction of the population of samples that will include all but a small percentage of the cancer cases. Alternatively, an instrument that could sort suspicious cells from each sample and present just these cells to a pathologist in place of the usual large sample of cells could greatly increase the efficiency of cytological examination. Such a device must measure the properties of some 100,000 cells in a few minutes and must be integrated into a practical method of cell collection and preparation. Several devices other than the RCS have been proposed for this purpose (Mellors and Silver, 1951; Bostrom et al., 1959; Ladinsky et al., 1964; Murray, 1962).

Caspersson (1950), Mellors et al. (1952), and others (Kamentsky et al., 1963) have shown that some cancer cells have a larger absorption in the ultraviolet at the maximum absorption of nucleic acids than do normal cervical cells. This larger absorption is probably due to an increase in RNA of the cancer cells with respect to other exfoliating cells. This enhanced nucleic-acid absorption is probably not specific, and other abnormalities may also produce enhanced absorption to varing degrees. However, our intial studies with the RCS indicated that measurement of absorption at 2600 Å offered a relatively simple screening criterion. Only machine tests could determine whether the screening efficiency of this modality was large enough to be of practical value (see below).

4. Cell Identification and Sorting

The practical need for multiple-wavelength measurements is illustrated in Fig. 4-7 where normal and cancer cervical cells photographed at 2600 Å are shown. The enhanced cytoplasmic absorption of the more undifferentiated cancer cells is apparent. However, the clump of normal epithelial cells has a total absorption as great as that of any of the cancer cells. The ultraviolet absorption measurement must be normalized with respect to some measure of total mass of the cell or cell clump under observation. Measurement of cell scatter near 4000 Å appeared to give the best estimate of cell mass for our purposes.

A second major consideration in cervical screening by ultraviolet absorption is the contribution of scatter to the 2600Å absorption discussed in Section III. We have found that the factor $a\Phi$ of Eq. (9) is very dependent on the suspending medium and can vary for cancer cells from a value about equivalent to the absorption due to nucleic acids to almost ten times as much. Thus, if uv absorption is displayed against scatter on the RCS, there will be little discrimination among cell types unless the scatter contribution is minimized. This is illustrated in Fig. 4-8 where display patterns of a sample containing metastatic keratinizing epidermoid carcinoma is shown measured suspended in a medium (the Carnoy fixative) in which the scatter contribution is great, and also in a medium (acetic acid) in which the scatter contribution is small enough so that differentiation among cell types is apparent.

The RCS has been utilized to measure samples obtained by both vaginal wash and cervical swab. Preliminary results are described by Kamentsky *et al.* (1965), and the use of these measurements to sort cervical cells is described in another paper by Kamentsky and Melamed (1967b). Figure 4-5 illustrates a result of the cell sorter with the ratio of 2600 Å absorption to scatter as the classification criterion.

An early version of the RCS was clinically evaluated by Koenig *et al.* (1968) in a double blind test with cervical swab and vaginal wash specimens from 1155 patients during a 9-month period in 1965–1966. RCS results were compared with concurrently obtained samples processed by the Papanicolaou technique and independently evaluated. Patients considered to have carcinoma of the cervix were biopsied. Only biopsy results were accepted as correct and final in evaluating the RCS. A major problem in these tests was obtaining adequate swab samples so that the required 100,000 cells per sample could be obtained. Only 45% of the samples were adequate at that time. This problem may be solved through the use of a new swab-collection technique. The RCS selected 34% of the readable samples, while including 85% of the readable cases with cancer. These tests reflected the efficiency of the cell-sampling procedures as well as the performance of the device itself and probably accounted for some of the

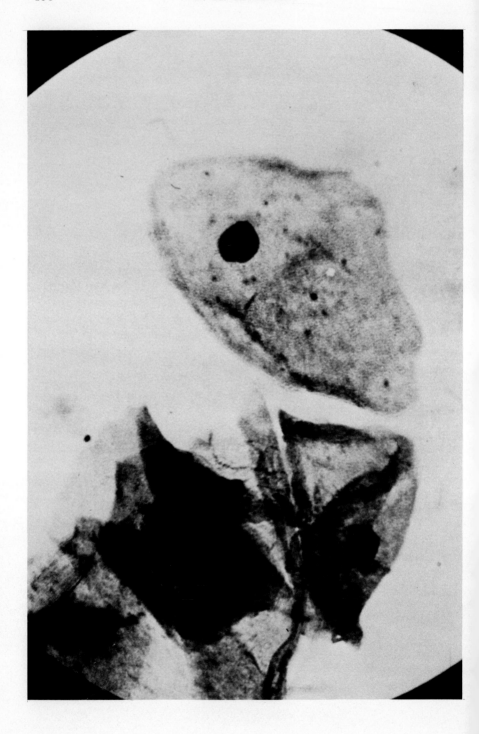

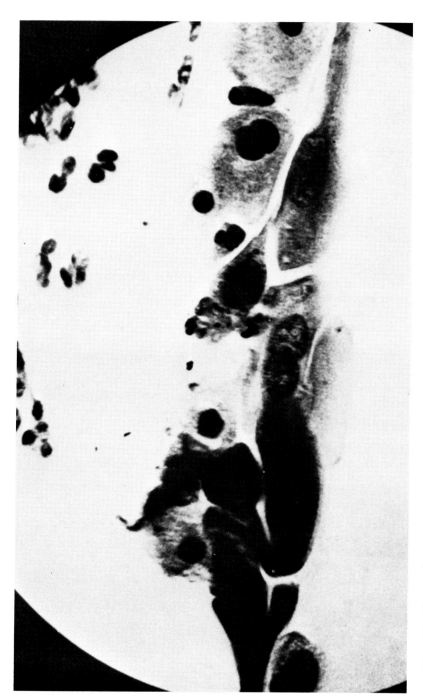

FIG. 4-7. Normal cervical cells (a) and cancer cells of varying degrees of differentiation (b) photographed at 2600 Å.

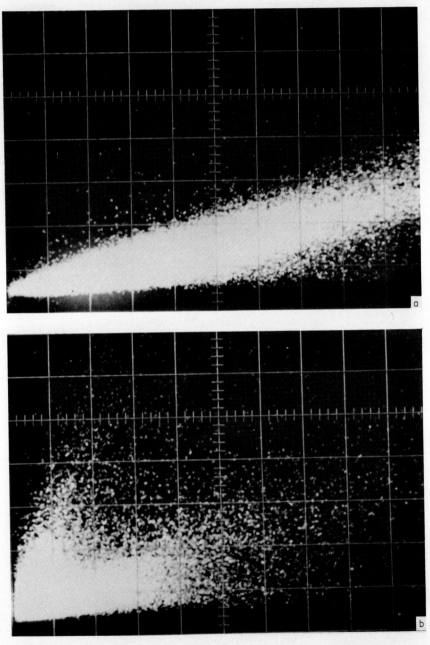

Fig. 4-8. RCS display patterns of metastatic keratinizing epidermoid carcinoma suspended in Carnoy fixative (a) and in acetic acid at pH 2.1 (b).

cases that were missed with the swab samples. The cytology miss rate during this test was substantially the same as with the RCS.

Sandritter *et al.* (1966), Caspersson (1964), and Weid *et al.* (1966) have recently made independent studies comparing the amounts of DNA, RNA, and cell mass in normal and cancer cervical cells. Their results indicate that the DNA content per cell may be a more specific indicator of carcinoma of the cervix than RNA. This result may be utilized with the RCS using absorption or fluorescent staining techniques. Currently, we are evaluating both fluorescent and absorbing metachromatic dyes for RNA and DNA in combination with a measurement of total protein by scatter or absorption of a protein stain.

B. Viability Assay

The determination of the relative numbers of viable and dead cells in a preparation is of increasing clinical importance with the development of surgical transplantation of organs. Two methods of utilizing the RCS to count the numbers of live and dead cells in a sample have been studied.

The first method is based on the use of the dye fluorescein diacetate as described by Celada and Rotman (1967). This dye is metabolized by viable cells and broken down to fluorescein, which is fluorescent at 5100 Å when excited by ultraviolet light. Dead cells will scatter light somewhat more than live cells. The RCS was set up to measure light emmission at 5100 Å simultaneously with cell scatter at 3660 Å. Figure 4-9 shows a typical display using mouse lymphocytes as a test sample. Classification in this case is obtained by simply counting signals above a threshold on each of the axes.

Since fluorescein diacetate reacts with serum used in inmunological studies, an alternative technique was developed to avoid washing the cells after incubation to remove the serum. The cells were stained with Trypan Blue, which is absorbed only by the dead cells. A dark-field system was used in the RCS and the condenser numerical aperture adjusted so that there was some bright-field illumination as well. In dark field the live cells scattered light well enough to provide an adequate emission signal. The Trypan Blue-stained cells presented enough absorption at 5900 Å to overcome the scatter signal and appear to the photodetector as absorption signals. Thus the live cells were found to emit light and the dead cells to absorb it. The positive and negative signals produced are independently counted to give the number of live and dead cells present in the sample. This technique has been evaluated by Melamed *et al.* (1968) with mouse lymphocytes, thymocytes, and other cells. Well-behaved serum titration

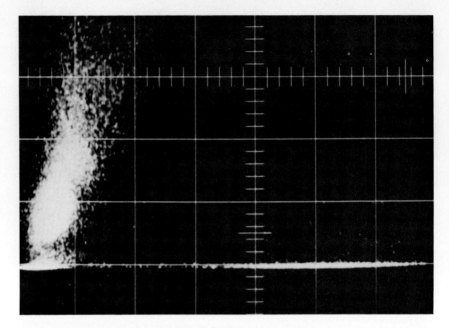

Fig. 4-9. RCS display pattern of mouse lymphocytes stained with fluorescein diacetate. The vertical coordinate is fluorescence at 5100 Å; the horizontal coordinate is scatter at 3660 Å.

curves were obtained with the RCS that corresponded with manual determinations of cell viability.

C. Differential Blood-Cell Count

Obtaining the relative numbers of the different blood-cell fractions by machine has great practical importance. As of this writing the RCS, through the use of a number of different staining techniques, has been used successfully to discriminate among three major blood-cell fractions, erythrocytes, lymphocytes, and granulocytes. O'Brien (1968) has described an application of a technique developed by Deitch (1955) using the stain Naphthol Yellow S for protein in conjunction with the Feulgen technique for DNA. Figure 4-3 shows a typical result using this technique.

The RCS is currently being evaluated with different methods of cell preparation and in conjunction with the 1130 computer to differentially count leukocytes to compare these counts with manual observations. Through the use of additional wavelengths to implement the techniques developed in Section III, we hope to identify the various granulocytes.

VI. SUMMARY

Devices like the RCS are currently at too early a state of development to evaluate their clinical use. The need to develop cell-identification instrumentation to perform tests on cells in the clinical laboratory in a similar automated manner to the way in which body fluids are analyzed by automated clinical laboratory instruments is great, however. The feasibility of the RCS for three cell identification applications is being studied.

There is also a great need for a preparative cell sorter. The principal of fluid switching developed for the RCS can be coupled to optical or electrical techniques for measuring cellular properties. The feasibility of cell sorting by fluid switching has been shown. However, the application of the RCS or devices like it to the clinical laboratory must still await engineering development.

Acknowledgments

I am pleased to acknowledge that all of the studies with the RCS described in this chapter represent joint efforts over the last four years with Myron R. Melamed, Herbert Derman, and Bo Thorell. Many people at IBM contributed to the development of the RCS instrumentation and programs. I am especially grateful to Isaac Klinger and George A. Folchi, who were responsible for much of the electronic and mechanical design, and to Kenneth W. Case, Jose I. Fortoul, and Garry H. Hatfield, who were responsible for the 1130 interface and programming support. This project has been generously supported by IBM, primarily through the efforts of Richard L. Garwin and Seymour H. Koenig.

References

Boddington, M. M., and R. A. Diamond. (1967). *Brit. Med. J.* **3**, 160.
Bonner, R. (1965). *Ann. N.Y. Acad. Sci.* **128**.
Bostrom, R. C., H. S. Sawyer, and W. E. Tolles. (1959). *Proc. IRE*, **47**, 1895.
Caspersson, O. (1964). *Acta Cytol.* **8**, 45.
Caspersson, T. (1950). "Cell Growth and Cell Function, a Cytochemical Study," Norton, New York.
Caspersson, T., G. Lomakka, and G. Svensson. (1957). *Exptl. Cell Res. Suppl.* **4**, 9.
Celada, F., and B. Rotman. (1967). Information Exchange Group No. 5 Scientific Memo No. 318.
Cornwall, J. B., and R. M. Davison. (1960). *JSI* **37**, 414.
Coulter, W. H. (1956). *Proc. Natl. Electron. Conf.* Chicago, Ill.
Deitch, A. D. (1955). *Lab. Invest.* **4**, 324.
Fulwyler, M. J. (1966). *Science* **150**, 910.
Gelerenter, H. L., L. A. Kamentsky, and R. J. Potter. (1966). U.S. Patent 3,270, 611.
Highleyman, W. H. (1962). *Proc. IRE* **50**, 1501.
Highleyman, W. H., and L. A. Kamentsky. (1959). *Proc. Western Joint Computer Conf. San Francisco 1959*, 291.

Kamentsky, L. A. (1959). *Proc. Western Joint Computer Conf. San Francisco 1959*, 304.
Kamentsky, L. A. (1961). *IRE Trans. Electron. Computers* **EC-10**, 489.
Kamentsky, L. A., and C. N. Liu. (1963). *IBM J. Res. Develop.* **7**, 2.
Kamentsky, L. A., and M. R. Melamed. (1967a). *Ann. N.Y. Acad. Sci.* Conference on Data Extraction and Processing Optical Images, New York City.
Kamentsky, L. A., and M. R. Melamed. (1967b). *Science* **156**, 1364.
Kamentsky, L. A., H. Derman, and M. R. Melamed. (1963). *Science* **142**, 1580.
Kamentsky, L. A., M. R. Melamed, and H. Derman. (1965). *Science* **150**, 630.
Koenig, S. H., R. D. Brown, L. A. Kamentsky, A. Sedlis, and M. R. Melamed. (1968). *Cancer* **21**, 1019.
Ladinsky, J. L., G. E. Sarto, and B. M. Peckham. (1964). *J. Lab. Clin. Med.* **64**, 970.
Melamed, M. R., L. A. Kamentsky, and E. A. Boyse, (1968). *Science* **163**, 285.
Mellors, R. C., and R. Silver. (1951). *Science* **114**, 356.
Mellors, R. C., J. F. Keane, Jr., and G. N. Papanicolaou. (1962). *Science* **116**, 265.
Mendelsohn, M. L. (1958). *J. Biophys. and Biochem. Cytol.* **4**, 407.
Moldavan, A. (1934). *Science* **80**, 188.
Montgomery, P. O'B. (1962). *Ann. N.Y. Acad. Sci.* **97**, 329.
Murray, N. A. (1962). *J. Florida Med. Assoc.* **48**, 1052.
O'Brien, R. T. (1968). *Intern. Cong. Histochem. Cytochem.* 3rd, New York.
Ornstein, L. (1952). *Lab. Invest.* **1**, 250.
Pateau, K. (1952). *Chomosoma* **5**, 341.
Potter, R. J. (1964). *SPIE J.* **2**, 75.
Prewitt, M. S., and M. L. Mendelsohn. (1965). *Ann. N.Y. Acad. Sci.* **128**, 1035.
Pruitt, J. C., A. W. Hilberg, R. F. Kaiser, S. C. Ingraham, II, S. J. Smith, and M. B. Willoughby. (1959). *J. Natl. Cancer Inst.* **22**, 110.
Sandritter, W., M. Carl, and W. Ritter (1966). *Acta. Cytol.* **10**, 26.
Sebestyen, G. S. (1962). "Decision Making Processes in Pattern Recognition." Macmillan, New York.
Sweet, R. G. (1963). *Electron. Design* Oct. 11.
Technicon Inst. Co. (1966). "The Technicon Cell Counter System" Technicon Inst. Co., Ardsley, N.Y.
Unger, S. H. (1958). *Proc. IRE* **46**, 1744.
Weid, G. L., A. M. Messing, and E. Rosenthal. (1966). *Acta. Cytol.* **10**, 31.

CHAPTER 5

Use of a High-Speed Digital Computer in Processing Radioisotope Scintiscan Matrices

W. NEWLON TAUXE

SECTION OF CLINICAL PATHOLOGY, MAYO CLINIC AND MAYO FOUNDATION, ROCHESTER, MINNESOTA

DONALD W. CHAAPEL

MEDICAL APPLICATIONS DIVISION, IBM CORPORATION, ROCHESTER, MINNESOTA

ALLAN C. SPRAU

APPLIED MATHEMATICS DIVISION, IBM CORPORATION, ROCHESTER, MINNESOTA

I.	Introduction	145
II.	Present Technique	146
	A. Source of Data	146
	B. Method of Data Collection	146
III.	Results	148
IV.	Conclusions	158
	References	158

I. INTRODUCTION

The human perceptive process is limited in its ability to distinguish various densities of gray tones at certain levels, to the point that much of the data contained in an ordinary photographic radioisotope scintiscan are lost. This is especially true when the scintiscan is presented in its usual *pointillistic* mode.

To assist human perception, various means have been sought to enhance the contrast between two adjacent but visually indistinguishable shades of gray. An important step was achieved in the presentation of radioisotope scintiscan data by electronic contrast-enhancement analog systems (Bender and Blau, 1960) and by color coding of various counting intensities (Kakehi et al., 1962). These mechanisms served principally to point up the nature of the problem, rather than to effect any significant solution of it.

The failure of these systems to achieve an effective solution to the problem is generally due to the fact that the data are in analog form. They cannot be manipulated mathematically or evaluated statistically. Analog systems are frequently difficult to modify and tend to be expensive. Handling of the data in the digital mode is not encumbered by these disadvantages. Because of the large amount of data generated on a scintiscan plot, a high-speed digital computer for manipulation and presentation of such data is extremely useful. This chapter presents our experiences with such a computer in the scanning operation. Descriptions of various stages of development of our program have appeared in the literature (Dolan and Tauxe, 1968; Sprau et al., 1966; Tauxe et al., 1966a; Tauxe et al., 1966b; Tauxe, 1968; Tauxe, 1969).

It should be stressed that our efforts have been concentrated on the midportion of the scanning process. No important modifications have been made on the probe itself; digitized x–y orientation signals have been generated and additional pulse-height analyzers have been added. At this point we consider our final readouts something less than ideal. The data-manipulation process has received our greatest attention.

II. PRESENT TECHNIQUE

A. Source of Data

Studies were made on data generated from scintiscans of patients referred to the radioisotope laboratory for routine diagnostic procedures. Occasionally, plastic phantoms containing radioactive materials were also scanned.

B. Method of Data Collection

All data derived during the process of scintiscanning were collected on an eight-channel digitizing magnetic tape recording system. Data were recorded transversely across the tape instead of by the usual lengthwise registration. The data-collecting channels were utilized as follows:

Channels 1 through 4, data from pulse-height analyzer.
Channel 5, body landmarks.

5. Use of a High-Speed Digital Computer 147

Channel 6, x orientation system.
Channel 7, y orientation system.
Channel 8, end of scan and end of series.

Also recorded on the tape were various patient-identification data such as patient's name, clinic registration number, date, organ to be scanned, isotope used, date of scan, projection of scan, and type of collimator system used. For this process the eight channels were used in a slightly different mode.

The tapes were fed into an IBM 7040 for processing. The computer was programmed to reconstitute the various counting rates in correct x–y orientation. Each datum point was then evaluated to correspond to the resolving power of the collimator system used. The entire matrix was searched for the point of maximal intensity; this value was divided into 100 increments of counting rate, each increment equalling 1% of maximal counting rate. These increments were then collected into groups of five for printouts. Thus, 20 levels were depicted on a single format. Various symbols of type were used to depict these counting-rate levels (these can be chosen at will). Those chosen for printouts for the examples given in this paper were as follows: I, blank, ., 0, blank, ,, W, blank, —, M, blank, +, 8, blank, =, @, blank, T, *, and blank.

Each one of these symbols could be used to represent a different group of five 1% counting-rate increments. For example, the I, which represents the lowest increment, may represent 0–5% on one scan readout, 1–6% on a second readout, 2–7% on the third, 3–8% on the fourth, and 4–9% on the fifth. Each of the other symbols would be shifted also by 1% to record the other levels. In this way, all 100 counting-rate increments can be depicted on five readouts. If these are printed on transparent paper and superposed in registry on each other, 100 levels can be seen at once without encumbering the eye. This is because only 20 symbols are seen, with each symbol representing five different intensity levels at five different gray tones produced by the superposition of matrices. Single counting-rate increments can be printed in other combinations or intensity groupings. Alternatively, an x–y plotter can be directed by the computer to draw the contour of each or of selected count-rate increments.

The scan matrix may be processed in other ways as well. The matrix derived from one pulse-height analyzer may be added to, subtracted from, multiplied by, or divided by the matrix derived from another pulse-height analyzer. Secondary matrices may be derived by making a matrix of the absolute differences between adjacent counting rates. We have referred to this as a differential plot. This matrix may be added to another matrix, which may bring out irregularities in the original matrix. All the counting rates on a given matrix may be summed, thus permitting total organ uptakes to be depicted, or sums of activities over rows or columns may be added and plotted.

Most of the scans presented here were registered against magnification-free roentgenograms made by the technique of Lewall and Tauxe (1967).

III. RESULTS

We have found such scan processing to be of value in the interpretation of scintiscans of all organs thus far tested. It has permitted a thyroid uptake to be determined from the scan itself. Ratio scans have permitted estimation of the distance of the source from the scanning probe. Also, computer processing has permitted interpretation of scans made at extremely low counting rates. It has improved resolution to the point that much smaller lesions can be depicted than are obvious by any other techniques when safe radioactivity levels are used.

Figure 5-1 shows (a) a photoscan, (b) the effect of simple digitization of the scan matrix by the computer, and (c) the effect of reprocessing after correcting the data in accordance with the resolution of the data-gathering mechanism.

Figure 5-2 illustrates a typical scan (a) and the variation of readout capabilities of computer-processed scans (b–e). These include (b) a 20-level computer-processed scan, (c) a 100-level plot derived from the same data (showing the fine detail possible in such representations) and (d) three-level plots from the 20-level plots (in which asterisks represent one specific 5% level, blanks represent all levels below this, and dashes represent all levels above it). It is possible to superpose 20 of such three-level plots that have been printed on transparent paper (e). This resembles a photoscan, but all levels have been corrected in accordance with the resolution capabilities of the data-gathering mechanism. It is difficult to distinguish individual levels. Figure 5-2f represents 20 of the counting-rate levels drawn by an x–y plotter. This has the advantage of being fairly easy to recognize, although relatively slower to produce; some variation of this method may represent the best mode of clinical reporting in the future.

Figure 5-3 illustrates the abilities of the system to handle scans of extremely low counting rates. Figure 5-3a is a photoscan of a thyroid after 5 μCi of ^{131}I had been administered to the patient instead of the usual dose of 50 to 100 μCi. Only 0.8 μCi was in the thyroid at the time of the scan. It is a poor scan, from which little information can be gathered. Figure 5-3b is a computer-produced contour plot of 20 levels of counting rates made from a tape recording of the same data used to produce the

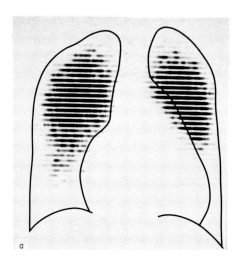

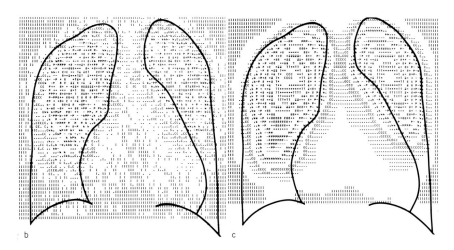

FIG. 5-1. Study of patient with mitral stenosis. (a) Photoscan of lung superposed on tracing of magnification-free chest roentgenogram. It is obvious that radioactivity is apically distributed. (b) Digital output of same scan, in which various characters have been assigned to represent various counting rates in correct x–y orientation. Distribution is still patchy and scan is difficult to comprehend. (c) Same scan as in (b) except that data have been processed to accord with resolution capabilities of collimator. (From Sprau, A. C., W. N. Tauxe, and D. W. Chaapel (1966). A computerized radioisotope-scan-data filter based on a system response to a point source. *Mayo Clin. Proc.* **4**, 585–598).

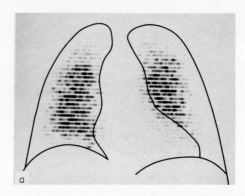
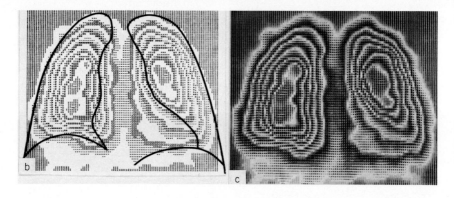

FIG. 5-2. Normal lung after injection of ^{131}I-labeled macroaggregated human serum albumin. (a) Photoscan. (From Tauxe, W. N. (1968). 100-level smoothed scintiscans processed and produced by a digital computer. *J. Nucl. Med.* **9**, 58–63. By permission of the Society of Nuclear Medicine.) (b) Computer-processed matrix in which 20 levels of activity are represented by various symbols. (From Tauxe, W. N. (1968). Über die Auswertung von Radioisotop-Szintigrammdaten durch Computer. *Meth. Inform. Med.* **7**, 96–104. By permission of F. K. Schattauer-Verlag.) (c) Computer-processed matrix in which 100 levels of activity are represented. (From Tauxe, W. N. (1968). 100-level smoothed scintiscans processed and produced by a digital computer. *J. Nucl. Med.* **9**, 58–63. By permission of the Society of Nuclear Medicine.) (d) Three three-level plots in which three different counting-rate increments (5% of maximum) are represented by asterisks. Counting rates below the increment depicted are represented by blanks, and those above it are represented by hyphens. (e) Superposition of 20 three-level plots. This results in an output similar to conventional plots. The eye has difficulty in differentiating individual intensity levels at the upper end of the scale. (f) Contour plot, by computer-driven x–y plotter, of 20 counting-rate levels. Each level is numbered for ease of recognition. (From Tauxe, W. N. (1968). Über die Auswertung von Radioisotop-Szintigrammdaten durch Computer. *Meth. Inform. Med.* **7**, 96–104. By permission of F. K. Schattauer-Verlag.)

5. Use of a High-Speed Digital Computer 151

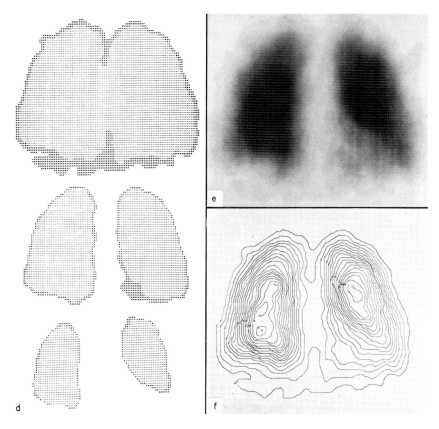

photoscan. Much more information is available on this plot. Not shown on this matrix but printed routinely on the drawings is a key to the counting rates represented by each level, the total counting rate present at each level, and the sum of all counting rates at any particular level and those above it. From this sum it is impossible to compute the total radioactivity over the thyroid gland and to determine the percentage uptake by the gland. Figure 5-3c illustrates a 100-level plot of this thyroid scan matrix. It obviously contains much more detail than the original photoscan. The point of maximal intensity over the right lower lobe corresponded with a hyperfunctioning nodule in the patient.

Perhaps the most graphic of all data representations so far attempted have been three-dimensional models made from the contour plots. Figure 5-4a is another photoscan of the thyroid. It is a poor representation and is difficult to interpret when seen alone. Figure 5-4b is an x–y plot of the scan data matrix; the various intensity levels are numbered in increasing order. The model constructed from this is shown in Figure

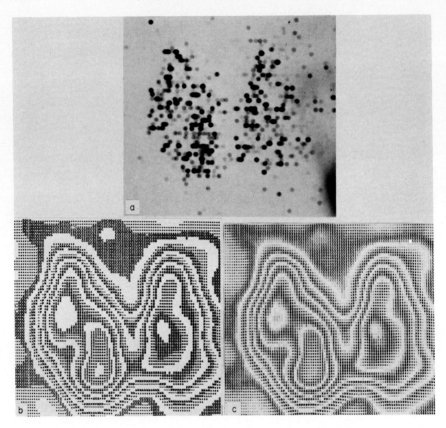

FIG. 5-3. (a) Photoscan of thyroid containing low level of activity—only 0.8 μCi of ^{131}I. (From Tauxe, W. N. (1968). 100-level smoothed scintiscans processed and produced by a digital computer. *J. Nucl. Med.* **9**, 58–63. By permission of the Society of Nuclear Medicine.) (b) A 20-level plot of data. (From Tauxe, W. N. (1968). Über die Auswertung von Radioisotop-Szintigrammdaten durch Computer. *Meth. Inform. Med.* **7**, 96–104. By permission of F. K. Schattauer-Verlag.) (c) A 100-level plot of this matrix. (From Tauxe, W. N. (1968). 100-level smoothed scintiscans processed and produced by a digital computer. *J. Nucl Med.* **9**, 58–63. By permission of the Society of Nuclear Medicine.)

5-4c. Here one can readily sense the presence of a nonfunctioning tumor in the right lobe of the thyroid, distorting that lobe by pushing it across the midline. This is not obvious from the photoscan (except in retrospect). However, at present, such models are impractical.

On numerous occasions we have observed lesions on computer-processed plots that were not evident on the routine photoscans. These have been seen in both positive tumor delineation (as in brain scans) or negative tumor delineation (as in liver scans).

5. Use of a High-Speed Digital Computer

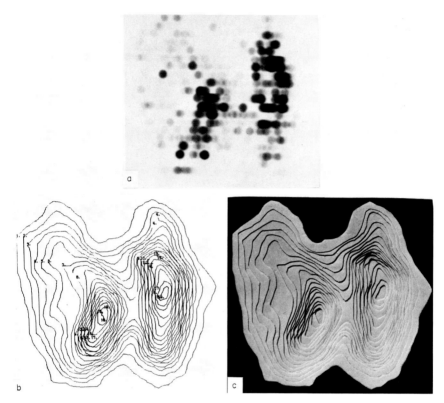

FIG. 5-4. (a) Conventional photoscan of thyroid, illustrating the difficulty in discerning patterns of distribution when data are shown in analog format. (b) Isocount contour plot of these data. (c) Three-dimensional model made from (b).

For example, a photoscan (Fig. 5-5a) of a brain that contained a cystic astrocytoma did not contain sufficient contrast to permit positive detection of the lesion. On the other hand, a computer-processed matrix (Fig. 5-5b) of this scintiscan revealed regions in the frontal lobe that were at least four increments (each increment 5% of maximal intensity) greater than the common level; these corresponded with solid areas of this tumor. In the parietal region there was an area four 5%-maximal-intensity increments less than the common level; this corresponded with a cystic area of the tumor.

A routine photoscan (Fig. 5-6a) representation of the distribution of ^{131}I-labeled Rose Bengal in the liver appeared completely normal and was so diagnosed until the computer plots (Fig. 5-6b) were seen. These revealed a distortion in the contour of the upper portion of the right lobe, which was not evident on the photoscan. This contour distortion

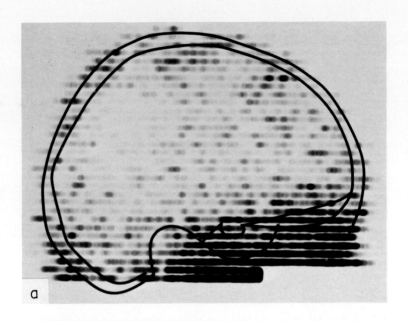

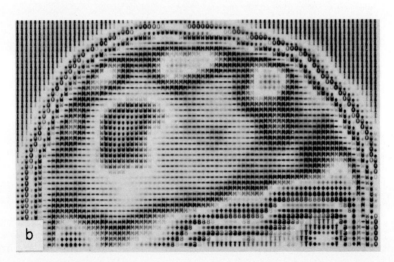

Fig. 5-5. (a) Conventional photoscan of brain, which was deemed negative in this format. (b) A 100-level plot of these data, which reveals areas in the frontal lobe that are at least four 5% increments above common level and an area in posterior parietal region that is at least four 5% increments below common level. The former was associated with solid and the latter with cystic areas of a high-grade astrocytoma.

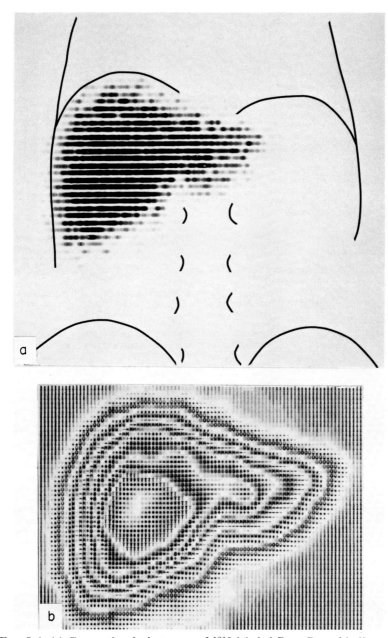

FIG. 5-6. (a) Conventional photoscan of ^{131}I-labeled Rose Bengal in liver; this appears normal. (b) A 100-level plot of these data reveals distortion in the contour patterns of the upper pole of the right lobe. This corresponded with a focus of metastatic hemangiopericytoma.

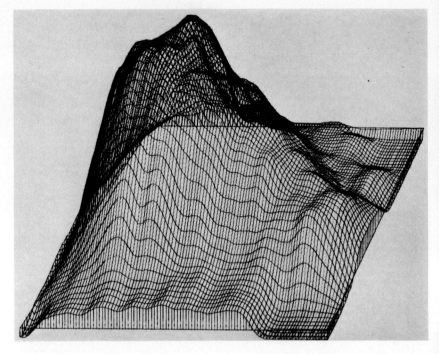

Fig. 5-7. Liver scan drawn by an *x–y* plotter under computer control, as it would appear from below.

corresponded with a focus of metastatic hemangiopericytoma proved by needle biopsy at the time of laparotomy.

We do not consider any of the readout formats particularly ideal for any given situation, and certainly none is ideal for all. The ones presented as type symbols have the advantages of speed and easy quantifiability, but they are somewhat difficult to interpret by the inexperienced eye. The contour plots are perhaps more readily recognizable, but it is not possible to discern individual levels unless they are labeled.

It has been our impression that a three-dimensional model provides the most information. While actual construction of such models in a clinical situation is impractical, the depiction of such by *x–y* plotter does not appear to be so. This requires that the matrices be depicted as they would appear if they were viewed from other angles. An example of this form of data presentation is shown in Fig. 5-7, an ^{131}I-labeled Rose Bengal scan of the liver as it would be seen from below. It has the advantage of permitting differentiation of relatively close counting rates, as might be evident on an actual model, but some areas are hidden. Multiple views are

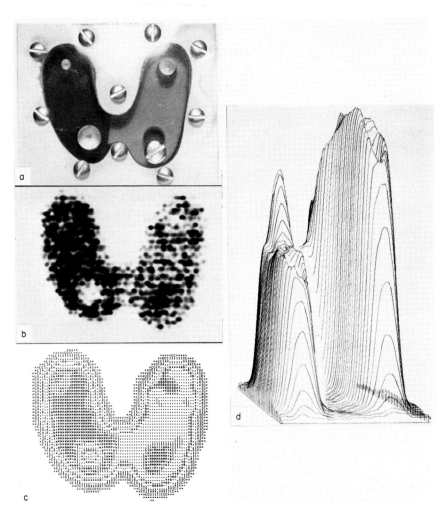

FIG. 5-8. (a) Plastic thyroid phantom containing dye solution to show three filling defects ("holes") and one well ("hot spot"). The left side has twice the depth of right. (b) Photo-scan of phantom filled with solution containing ^{131}I. Only that the right lobe is more intensely radioactive than the left and that a filling defect is clearly discernible at the right lower pole can be deduced clearly. (From Sprau, A. C., W. N. Tauxe, and D. W. Chaapel: A computerized radio-isotope-scan-data filter based on a system response to a point source. (1966.) Mayo Clin. Proc **41**, 585–598.) (c) Computer-processed plot. All filling defects are discernible, and the total activity over the "gland" may be ascertained. (From Sprau, A. C., W. N. Tauxe, and D. W. Chaapel. (1966). A computerized radioisotope-scan-data filter based on a system response to a point source. *Mayo Clin. Proc.* **41**, 585–598.) (d) An x–y plot of these date in which the "gland" is seen as if from over the left shoulder of patient.

necessary. This drawing can also be produced in different colors to be viewed through colored lenses for a three-dimensional effect.

Figure 5-8a depicts representations of a plastic thyroid phantom (Picker NI-94198) filled with ^{131}I. Three "defects" are present, two in the left lobe and one in the right. The left lobe is twice as thick as the right. There is also a well in the lower right pole, partially covered by a screw used as a plug. Figure 5-8b depicts a routine photoscan of this phantom, in which it is easy to discern that the right lobe is different from the left. The 11-mm plug in the right lower pole is easily seen, but the other defects are not clearly seen in this analog format. After computer processing, all of the defects—both "holes" and "hot spots"—are clearly seen (Fig. 5-8c), along with the fact that the ledge between the two lobes is at an angle. The distance between this ledge and the nearby 11-mm plug is only 3 mm, and it appears well resolved. This presentation still possesses the disadvantage of not being completely interpretable without a key to the counting rates; a key is usually typed below. When coupled with a contour plot (Fig. 5-8d), however, the data become comprehensible. Perhaps some such coupling will become the ideal scan representation in the future.

IV. CONCLUSIONS

In general, our experience with computer processing of scintiscan matrices has indicated that it possesses many advantages over conventional means: (1) it provides increased resolution; (2) it handles low counting rates well, so that dosages of radioactive tracers can be decreased in many cases; (3) its cost will ultimately be lower than that of currently used methods; and (4) it will allow one to quantify scan data in ways not possible by the currently used analog methods.

References

Bender, M. A., and M. Blau. (1960). *J. Nucl. Med.* **1**, 21.
Dolan, C. T., and W. N. Tauxe. (1968). *Am. J. Clin. Pathol.* **50**, 83.
Kakehi, H., M. Armimizu, and G. Uchiyama. (1962). *In* Progress in medical radioisotope scanning. *Proc. Symp. Med. Div. Oak Ridge Instit. of Nuclear Studies, October 22–26, 1962"* (R. M. Knisely, G. A. Andrews, and C. C. Harris, eds.), pp. 111–131. Oak Ridge Institute of Nuclear Studies, Oak Ridge, Tennessee.
Lewall, D. B., and W. N. Tauxe. (1967). *Am. J. Clin. Pathol.* **48**, 568.
Sprau, A. C., W. N. Tauxe, and D. W. Chaapel. (1966). *Mayo Clin. Proc.* **41**, 585.
Tauxe, W. N. (1968). *J. Nucl. Med.* **9**, 58.
Tauxe, W. N. (1969). *J. Nucl. Med.* **10**, 258.
Tauxe, W. N., H. B. Burchell, D. W. Chaapel, and A. Sprau. (1966a). *J. Appl. Physiol.* **21**, 1381.
Tauxe, W. N., D. W. Chaapel, and A. C. Sprau. (1966b). *J. Nucl. Med.* **7**, 647

CHAPTER 6

Radiation Treatment Planning Using a Display-Oriented Small Computer

J. R. CUNNINGHAM and J. MILAN[1]

THE ONTARIO CANCER INSTITUTE, TORONTO, CANADA

I.	Introduction	159
II.	General Description	161
	A. Graphical Input	161
	B. Graphical Output	161
	C. Digital Input and Output...........................	163
	D. The Central Processing Unit	163
III.	The Radiotherapy Environment	163
IV.	Data Input	166
	A. Patient Contour	166
	B. Beam Data	166
V.	Graphic Output..	168
VI.	Data Manipulation	169
	A. Correction and Summation	169
	B. Hard Copy	171
VII.	Evaluation	171
	A. Accuracy..	171
	B. Usefulness	175
	References	178

I. INTRODUCTION

Radiotherapy is the use of ionizing radiation for the treatment of patients suffering from cancer. Very frequently the tumor to be eradicated is either deep within the patient or lies close to structures that would be damaged by radiation. Consequently, one of the chief problems in radiotherapy is to supply sufficient radiation at the site of the tumor to destroy it without at the same time harming the healthy tissues near by. The

[1] Present address: Department of Physics, The Royal Marsden Hospital, Sutton, Surrey, England.

determination of the optimum arrangement of one or several radiation beams and the calculation of the resultant dosage pattern is frequently called radiation treatment planning, although it is clear that a large number of other factors both physical and biological must be considered in planning a patient's treatment.

The calculation of a radiation-dosage distribution may be done by the radiotherapist himself, by his residents, by the physics staff, or by a well-trained technician. No matter who does it, it is a time consuming and tedious task. When done manually, it may take several hours' work to produce a radiation distribution, and then the distribution may or may not be acceptable for use. This is particularly true if the distribution is for a novel and sophisticated beam arrangement. Further, there is always a danger that, because of the labor involved, a plan may be judged acceptable even if it is not exactly what the radiotherapist might consider as best.

If it were possible to produce dose-distribution diagrams simply and quickly, it should be more practical to produce treatment plans that the therapist may consider optimum, even to consider more closely what would in fact be optimum. For a number of years large computers have been applied to the problem of dosage calculations (IAEA, 1966). The cost of such machines is great, and it is of course necessary to share them with many other users. This has at least two disadvantages. One is turn-around time, and the other is the difficulty of handling input data in graphical form. Turn-around time is frequently a day or more, and this is not short enough to allow multiple trials to be made if the patient is waiting. Time sharing might well be considered an answer to the turn-around problem, but virtually all time-sharing terminals readily available require a digital input from a keyboard type of device. A large part of the information that must be processed in treatment planning is in an analog form. An example of this is the outline of the body contour of the patient and the location and extent of the tumor. The digitization of this information requires a considerable amount of operator time. It is particularly desirable that the output be available also in graphical form, and although this facility is currently available with a few time-sharing systems, either it is expensive or the plotting time involved is longer than would encourage trials of alternative beam arrangements.

With these problems and a number of others in mind, a group at Washington University Biomedical Computer Laboratory, St. Louis, decided to explore the possibility of developing a small computer for radiotherapy treatment planning. It was to be a minimum-hardware computer oriented around specially designed input and output devices, easy to operate and fast enough that the operator could examine a number of radiation-dose distributions at a single sitting. It was also to be

6. Radiation Treatment Planning

inexpensive enough to become a potential acquisition for an average radiotherapy department. The result of this project was the Programmed Console. It was designed to operate both as a small independent computer, meeting the requirements discussed above, and as a remote terminal connected to a larger general-purpose machine. A discussion of its use as a small independent computer forms the substance of this chapter.

II. GENERAL DESCRIPTION

A. Graphical Input

The Programmed Console was designed with two basic principles in mind—special adaptation to graphical input and output, and minimum cost. The special devices for input are shown in Fig. 6-1. On the right can be seen a position transducer called the rho–theta device. It consists of a sliding arm mounted on a pivot. On the end of the arm is mounted a pointer that can be made to trace around a contour drawn on a piece of paper as seen in the photograph. The arm and the pivot are each connected directly to potentiometers to enable the production of voltages that are proportional to polar coordinates of the position of the pointer. The analog-to-digital conversion required to make these two signals compatible with the digital operation of the computer is performed primarily by means of programming rather than circuitry. This means that the computer must deal with the analog signals sequentially and does so by means of a programmable multiplexer.

In addition to the two voltages representing the polar coordinates of the rho–theta device, there is need for analog signals to control such parameters as the position and angle of incidence on a patient of a radiation beam. These signals are provided by potentiometers mounted at the top rear of the keyboard control shown on the left of Fig. 6-1. Up to four of these analog channels of input can be handled by the computer.

B. Graphical Output

Immediate graphical output is provided by means of an oscilloscope that has the facility to operate in a storage mode. This eliminates the complex hardware, or extensive programming, that would be required for continuous regeneration of the image. The image may be photographed, and this provides a rapid means of obtaining hard copy. Frequently a full-sized paper copy of the output is also required; this is obtainable by means of an incremental plotter which is readily adaptable to this system.

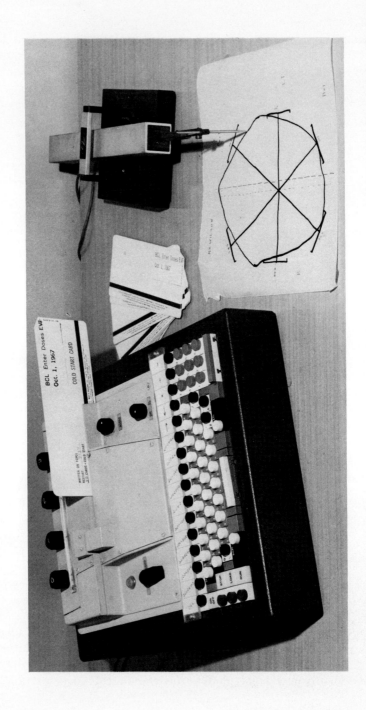

Fig. 6-1. The input devices of the Programmed Console. On the right can be seen the rho-theta device with its pointer about to trace around the contour of a patient's head. On the left can be seen the keyboard and the data master. Four knobs on the top of the data master provide analog input additional to the rho-theta device.

C. Digital Input and Output

Provision for input of digital information and mass storage is provided by a device that has been called the data master. This can be seen on the left in Fig. 6-1. This device operates using a card approximately 10 in. long on which is fastened a strip of magnetic tape. This can be seen in the diagram. The card carrying the tape can be fed at a constant rate past a reading or recording head. Approximately 256 words of information can be stored on the magnetic tape of the card. This device is about 10 times as fast as a teletype paper-tape system and removes the complications of controlling a magnetic-tape system. This is a convenient method for storing both programs and data. Additional input of digital information may be provided by the keyboard shown also in Fig. 6-1. This keyboard includes alphabetical and numerical characters as well as a number of additional keys that allow the computer to be operated completely from this keyboard. Digital output is obtained via a standard teletype unit readily available.

D. The Central Processing Unit

The central processing unit is built around a 4-K high-speed (2 μsec cycle time) 12-bit memory unit. The amount of hardware has been kept to the practical minimum, and in many ways its design is very similar to that of the LINC series of computers (Clark and Molnar, 1965).

III. THE RADIOTHERAPY ENVIRONMENT

Figure 6-2a shows a representation of the dosage pattern produced by a beam from a radiation-therapy unit. The beam is one from a cobalt-60 unit directed vertically downward on the plane surface of a tank of water. The curves shown in this diagram are isodose lines. Such isodose patterns are usually obtained directly, using an isodose plotter. This is a device that moves an ionization chamber around in a water tank in such a way that the desired isodose line can be "searched" out and its position plotted directly on paper (Johns, 1962).

The shape and spacing of the isodose lines depend on a number of factors, some of which are indicated in Fig. 6-3. The energy of the radiation, the cross-sectional area (field size) and the distance F from the source will affect the depth spacing of the lines. The diameter of the radiation source, the distance F_c between the source and the beam-defining diaphragm, and the distance F will all affect the shape of the isodose lines near the edge of the beam. Radiation scattered within the irradiated medium, as in

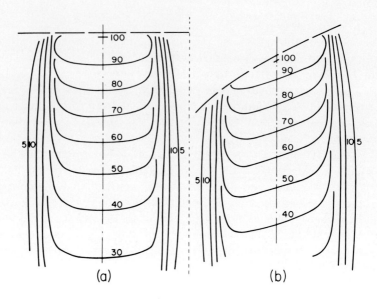

Fig. 6-2. (a) An isodose chart for a 10 × 10 cm beam from a cobalt-60 unit. This dosage distribution was obtained by means of ionization-chamber measurements made in a tank of water. (b) The dosage distribution of (a) has been altered to allow for a curved surface.

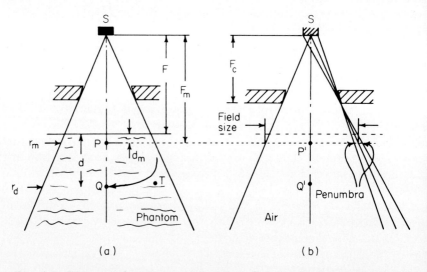

Fig. 6-3. Diagrams showing parameters that affect the shape of isodose distributions. The beam in (a) is irradiating an absorbing medium, while the one in (b) is in air.

6. Radiation Treatment Planning

Fig. 6-3a, will affect the general curvature of the isodose lines within the beam and will govern the position of the isodose lines outside the main beam. In addition to these factors, high-energy machines have beam-flattening filters that individually shape the beam. Because of these facts, there can be no completely general way of computing data for isodose curves, and therefore great reliance must be placed on measurements made with the beam parameters varying over the range that will be met in practice.

Manufacturers of radiotherapy equipment frequently supply isodose curves for their machines. The International Atomic Energy Agency has published an atlas of radiation dose distributions for single beams (1965) containing some 155 measured isodose charts.

Figure 6-2b shows the isodose pattern of Fig. 6-2a altered by a sloping and curved surface of a patient. It is clear that Fig. 6-2b cannot be measured directly (within the patient) but must be calculated from data such as shown in Fig. 6-2a. Methods for carrying out such a calculation are presented in ICRU report (1962) and a number of textbooks on the physics of radiology (Johns, 1962; Johns and Cunningham, 1969).

In radiation treatment planning, we wish to apply a number of such beams as are depicted in Fig. 6-2a to a patient contour and obtain a composite distribution from the summation of the dosage. An example of the result of three such beams applied to a patient contour is shown in Fig. 6-4. This distribution was obtained by manual calculation following methods outlined by Johns (1962). Such distributions have been produced in large numbers in large radiotherapy centers for at least a quarter of a century. Very many man hours have been spent at this task. In the distribution shown in Fig. 6-4, no correction has been made for the effect of the shape of the contour on each beam, although, as mentioned, this could have been included in the procedure. The grid of points seen in Fig 6-4 will be discussed in a later section.

It is first desirable that the computer follow procedures similar to those already followed by manual methods. To do so, a computer system to be useful in radiation treatment planning is presented with the following problems:

(1) *Data input.* The computer must be supplied with input data in a digital form.

(2) *Output.* Graphic output of the desired forms must be produced.

(3) *Data manipulation.* The necessary computation between input and output must be performed.

The specific way in which the Programmed Console has been applied to the solution of these problems, in the radiotherapy environment, will now be discussed.

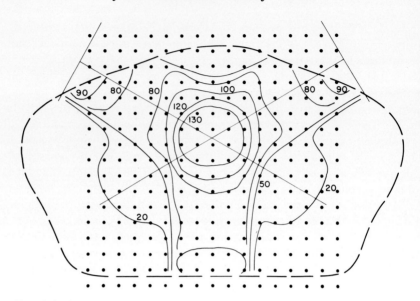

Fig. 6-4. Composite radiation-dosage distribution showing the results of combining three beams. Correction for body curvature has not been made. The grid superposed on this diagram represents points at which dose summations are carried out by the Programmed Console.

IV. DATA INPUT

A. Patient Contour

The digitization of the patient contour is accomplished by means of the rho–theta device and the "Enter Patient Contour" program. This program is stored on 15 data-master cards of the type seen in Fig. 6-1. As the outline, drawn on a piece of paper, is traced by the pointer of the rho–theta device, the computer is supplied with a series of polar coordinates. These are converted into cartesian coordinates by the program. Since this information must in turn be recorded on a Data Master card, the number of points that may be stored is limited by the card capacity. The separation of the points along the patient contour may be specified in order to balance precision with room for data storage. As desired, internal structures may also be traced.

B. Beam Data

In Fig. 6-5 the isodose chart of Fig. 6-2 has been redrawn with a grid of intersecting "fan lines" superposed on it. A digital value for the dose at the

6. Radiation Treatment Planning

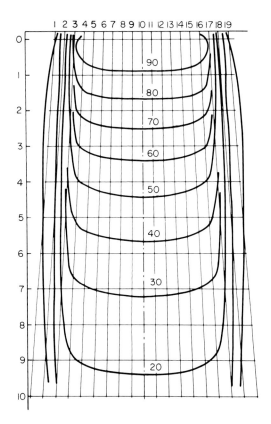

FIG. 6-5. A grid of "fan lines" is shown superposed on the isodose distribution of Fig. 6-2a. The magnitude of the dose must be obtained at each of the grid points. The fan lines, if extended upward, would converge at the radiation source.

intersections of each of these lines must be obtained. The fan lines are directed toward the radiation source to facilitate subsequent correction for patient contour. The position and number of the fan lines (normally 19) and the horizontal lines (normally 13) may be varied to obtain the most accurate description of a beam. The dosage distribution in the beam is thus described by a table of up to 256 numbers (normally 247) and can readily be accommodated on one data-master card along with a few other descriptive numbers such as the grid spacings, the beam size, and a title.

Any one of three methods may be used to obtain the dose table.

1. MANUAL DIGITIZATION. A grid such as the one shown in Fig. 6-5 may be superposed on a standard isodose chart and the numerical values obtained by interpolation from the isodose lines. A program "Enter Doses"

has been prepared to permit direct entry and editing of such data via the keyboard, displaying the numbers making up the dose table in the process. This method is the most laborious but remains the one of choice for beams that have had beam-shaping filters added to them for special purposes. The accuracy of this method is limited only by the patience of the operator.

A slight variation of this manual digitization is the production of the dose-table data via a larger computer. The numbers could then be entered manually as described above, or alternatively the Programmed Console could be used for this purpose as a remote terminal to a large central processor.

2. THE RHO–THETA DEVICE. An attempt has been made to produce a program so that the rho–theta device could be used to trace along isodose curves such as those shown in Fig. 6-2a to produce the beam table directly. This program is not as yet fully operative.

3. COMPUTATION OF BEAM TABLE. A program called "Beam Generator" has been written by one of the authors (JM) which computes the dosage values for the beam table from standard tissue–air ratio data. The method has been described in detail elsewhere (Johns and Cunningham, 1969) and will be discussed only briefly here. It is based on the separation of the radiation beam into primary and scattered radiation. The primary radiation can be handled either analytically or by a very small number of tabular data. Data pertaining to the scattered radiation are stored in the form of a table of scatter–air ratios (Gupta and Cunningham, 1966), which depend only on the energy of the radiation. The dose at any point in any shaped beam may then be computed by adding to the primary radiation a set of terms that are functions of the boundary conditions. Computations are valid over a very wide range of the parameters shown in Fig. 6-3, and allowance can be made for beam-shaping filters. The only input data required are the quantities F, the beam dimensions, and a set of numbers representing the radiation profile in air. These latter data allow for the actual beam penumbra and the effect of collimator and filters. They may be obtained by sweeping an ion chamber across the beam in air.

V. GRAPHIC OUTPUT

To obtain a resultant set of isodose curves from a chosen combination of beams it is necessary to determine a table of summated doses on a grid having equal x and y spacings between the lines. Such a grid is shown superposed on the isodose distribution of Fig. 6-4. The table may then be searched for specified doses, and this enables the corresponding points to be displayed on the oscilloscope. The magnitude of the spacing between

6. Radiation Treatment Planning

the points on the grid will determine the accuracy of positioning of the isodose lines but the size of the memory of the computer limits the number of points available. In the case of the Programmed Console the table has been chosen to contain $17 \times 17 = 289$ numbers. It is clear that a compromise must be made between accuracy and the size of the grid. As a solution to this problem, the grid may be made movable and variable in size. It may be large for rapid, rough assessment purposes, but small for viewing selected areas with higher accuracy.

VI. DATA MANIPULATION

A. Correction and Summation

A program that will produce isodose curves in the desired manner, known as the "Superimpose Beams" program, was developed. This program is contained on 12 data-master cards and when resident in the computer may be controlled by the keyboard and the four potentiometer knobs (analog channels) seen on top of the data master in Fig. 6-1. The upper row of keys are titled as follows; their functions describe the operation of the program.

(a) *Contour.* This key starts the data-master motor and enables the previously prepared card containing the patient contour to be read in.

(b) *New beam.* Enables a card containing beam data to be entered. The program will accept a maximum of three such cards.

(c) *Copy beam.* Enables the program to use beam data already entered as in (b) to add up to three additional beams to a maximum of six.

(d) *Weights.* Enables the intensity of each beam, including the copied beams, to be varied at will.

(e) *Plan.* Displays the patient contour together with the "viewing window" and bars representing the position of each of the beams as seen in Fig. 6-6. The viewing window corresponds to the boundaries of the square grid shown in Fig. 6-4. The size and position may be varied by means of data-master knobs after depressing key "0." As each beam is entered it is given a number from 1 to 6; the position and angle at which each beam enters the patient may be changed by means of the data-master knobs after depressing the number key corresponding to that beam. Individual beams may be removed from the display and from calculation by striking the number of the beam and then key "Off." It may be returned by depressing the number for that beam again.

(f) *Isodoses.* Depressing this key causes the computer to calculate and summate the doses at each point on the square grid that was shown in Fig. 6-4. To do this, it must deal with each beam sequentially. It calculates

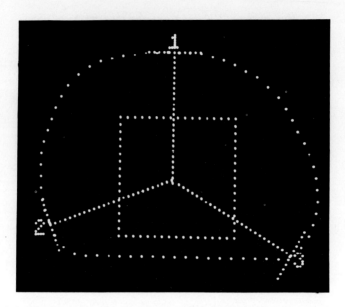

Fig. 6-6. Photograph of oscilloscope display produced by depressing key "Plan." The patient contour, as entered by the "Enter Contour" program, may be seen along with "T bars" indicating position and angle of three beams. The square represents the viewing window.

the position of each point on the square grid relative to the angle and position of entry of the beam to establish which fan and cross lines (Fig. 6-5) it has to inspect to determine the correct dose. Having established these, it determines the discrepancy between the contour (along that fan line) and the plane surface for which the beam table is valid and modifies the doses to allow for this discrepancy. The total dose to each grid point is found by summating the contributions from each of the beams.

Having finished the above calculation, the oscilloscope goes into the storage mode and displays the "viewing window," filling the screen with such parts of the contour as are contained within the window. Isodose lines may now be displayed on demand by entering the value for the desired isodose line from the keyboard. Up to 10 lines may be displayed at one time.

(g) *Settings.* Depressing the key "Settings" allows a number of subprograms to be brought into play. A data-master card is called for. If the Settings card is read in, data tables are displayed that describe the beams in use. The tables include the x and y coordinates of the beam entry point, the angle of entry, the relative beam intensity, and such items as beam size and title. Because of space limitation the subprogram is automatically obliterated after depressing key "Isodoses."

An alternative subprogram is "Multi Display." This program causes the area enclosed by the viewing window to be divided into 16 parts, after which the isodose curves are obtained in fine detail for each of the sixteen areas in turn. They are all displayed on the oscilloscope in the storage mode. This, in effect, multiplies the number of grid points by 16. The accuracy is increased, but at the expense of time and flexibility. The computer may be returned to normal operation by again depressing key "Settings" and entering the program card "Single Display."

B. Hard Copy

A permanent record may be obtained either by photographing the oscilloscope screen or by using an incremental plotter. For the latter, two additional programs must be read in following program "Superimpose Beams." One program, "Plot Isodose," consisting of 5 cards, causes the plotter to draw the previously determined isodose pattern in the "multi display" manner. The second, "Plot Plan," draws the patient contour on the isodose pattern and also prints the information on the beams that would be obtained by the "Settings" subprogram (Section VI,A).

VII. EVALUATION

There are two facets to the evaluation of the computer system just described—accuracy and usefulness. Accuracy must be at least as good as that now obtainable with manual methods, but greater accuracy would be desirable. Usefulness can only be judged by objective trial, with the full realization that the very presence of a computer system such as this will bring about changes in routines and techniques. The usefulness of these changes must also be assessed.

A. Accuracy

Accurate data for input should in principle be obtainable from existing isodose curves. The method of manual digitization should then be the most accurate of all. Tempering this slightly, however, is the fact that isodose curves, as they are now found in the literature (IAEA, 1965, for example), were not drawn with this particular use in mind. Although their accuracy in the main part of the beam should be beyond reproach, the position and shape of the isodose lines at the edges of the beam, i.e., in the penumbral region, is subject to errors. These may be caused by the size of the ion chamber used for the measurements, the accuracy with which the position of the ion chamber was reproduced by the plotting pen if an isodose plotter was used, and last but not least, the faithfulness by which the artist has reproduced the original data for publication. One of the authors (JRC) noted [during the manual digitization of isodose charts for input for the computation of moving field isodose charts (IAEA, 1967)] that very frequently isodose curves even for cobalt-60 beams were asymmetric. The disagreement in the shape of isodose lines between one side of the beam and the other was often 2–3 mm or more; furthermore, consistency was absent. The cause must have been one or all of the sources of error mentioned above. Most charts examined were not this bad, but because of the above factors great care must be exercised in assessing the accuracy of a computing system such as this by using isodose charts.

It is essential that the performance of the system be examined step by step. As a first step, manual digitization of a number of beams, both large and small, was carried out, and the output was compared directly to the original charts. It was found from this test that the viewing window grid that was originally used was too coarse. The linear interpolation procedure used to find the location of the isodose lines between the grid points was causing a large spatial distortion of the isodose lines. The "multi display" subprogram mentioned in Section VI,A was written to correct this and has done so for all practical purposes.

Since manual digitization and keyboard entry is itself a very tedious procedure, it was early decided that a program should be written to generate beam data and produce directly data-master cards carrying the beam formation. This program, "Beam Generator," was discussed briefly in Section IV,B,3. Figure 6-7 shows the plotter output of a 10×10 cm beam calculated for cobalt 60 for a source-to-skin distance (F of Fig. 6-3) of 80 cm. The dashed lines on this diagram have been traced from an isodose chart for this type of beam obtained by means of an isodose plotter.

The numbers shown labeling the isodose curves give the percentage dose relative to the maximum dose, which for this beam occurs on the central axis, 0.5 cm below the surface. The difference in dosage in the main part of the beam, as indicated by the two sets of curves, is not more than 1% of the maximum dose. The absolute error that would result, however, depends on the dose at a specific point. For example, a discrepancy of 1% at the 40% level would result in a relative error in dosage of $2\frac{1}{2}$%. Discrepancies in dosage at the edges of the beam are considerably greater.

Data pertaining to the edge of the beam are shown plotted in Fig. 6-8. The dotted curve, labeled "measured data," was obtained from the measured isodose chart (Fig. 6-5) by determining the positions of the intersection of the isodose lines with a horizontal line at a depth of 3 cm. The isodose chart gives the positions of doses at intervals of 10% only, and

6. Radiation Treatment Planning

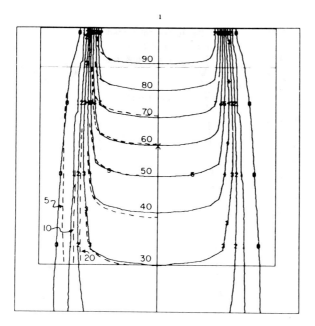

FIG. 6-7. Plotter-produced diagram for a 10 × 10 cm cobalt-60 beam calculated by program Beam Generator. The dashed lines are traced from an experimentally determined isodose chart (Fig. 6-2a) for the same beam. The dashed horizontal line marks a depth of 3 cm.

therefore the shape of the curve is uncertain between these values. The solid circles represent digital values produced by the "Beam Generator" program. These fit the experimental data very well over the central portion of the beam, but, in the region of the edge, deviate from it toward the beam center by as much as 3 mm. The open circles were taken from the plotted output and therefore represent "Beam Generator" input data that have been processed by the "Superimpose Beams" program. A solid line has been drawn through these points. It can be seen that the processing procedure, i.e., the double interpolation, from the fan lines to the grid points and from the grid points to the isodose lines has distorted the input data in such a way as to enlarge the penumbral regions of the beam. The amount of distortion will depend on both the fan-line spacing and the viewing-window grid spacing. Hence the distortion will be less for small beams by virtue of their small fan-line spacing, and will be less when small areas are being examined. The manifestation of this effect is the horizontal displacement of the output dosage values relative to the input (open circles versus solid circles in Fig. 6-8). This causes a dosage discrepancy of as much as 10% over a very small region.

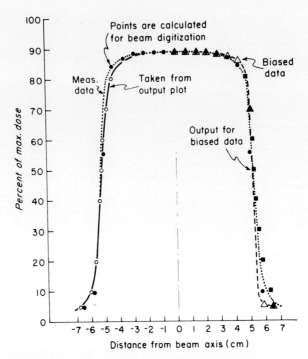

FIG. 6-8. Graphs showing radiation-dosage profile in a 10 × 10 cm cobalt beam at a depth of 3 cm.

The data shown on the left-hand side of Fig. 6-8 suggest that if one desires the output to more closely conform to measured isodose curves, the input data should perhaps be purposely adjusted to overestimate the sharpness of the beam. To test this, the 10 × 10 cm beam used for Fig. 6-8 was recalculated using an extremely small geometrical penumbra (see Fig. 6-3b). The results of this calculation are shown plotted as the triangles on the right-hand side of Fig. 6-8. It can be seen that a very sharp beam has been produced. The squares on this side of the figure are the points obtained from the display of these data after having been processed by the Superimpose Beams program. The dotted curve on the right is from the experimental isodose chart. The agreement between output and measurement can now be seen to be very good.

The authors are aware of the general inadvisability of purposely altering, and thereby decreasing the accuracy of, input data in an attempt to compensate for errors in data processing and do not recommend this procedure. It is unnecessary for beams with anything but very small penumbral regions. Furthermore, the Superimpose Beams program is currently being

6. Radiation Treatment Planning

modified to accept beam-data tables having an uneven fan-line spacing so that the spacing may be decreased over the edge of the beam. This enables far better description of the penumbral region and is a much sounder solution.

The accuracy of the correction made to each beam for oblique incidence and surface curvature is not discussed here. The method used is the effective SSD method and is discussed in ICRU Report 10d (ICRU, 1962, p. 23) and in standard textbooks dealing with the physics of radiotherapy such as that by Johns (1962) and Johns Cunningham (1969).

There is a natural tendency to compare computer-produced radiation distributions with those produced by hand on the grounds that the latter are more accurate. In the authors' opinion the comparison should be the other way around. Far more shortcuts are taken in the manual calculation than with the computer.

B. Usefulness

The first usefulness of the small, display-oriented computer lies in its ability to do rapidly and accurately what is very tedious to do by hand. In the authors' institution, the Programmed Console appears to be able to fit already established routines of treatment planning, and therefore a discussion of its usefulness would be largely a discussion of the usefulness of treatment planning.

Two small examples of the use of the system will now be presented however, as illustrations. Figure 6-9 shows a composite distribution that was obtained for the treatment of a tumor situated just lateral to the midline of a patient's chest. The spinal cord is not far behind the tumor, and the lungs are on either side of it. These areas should receive as low a dose as possible. A single beam from either front or rear, or a parallel and opposed pair of beams would spare the lungs, but the dose to the spine would actually be greater than that to the tumor. A distribution for three symmetrically arranged beams similar to the arrangement shown in Fig. 6-6 was next produced. It was noted from this that in order to keep the dose to the spine below about 60% of the tumor dose, the angulation of the two posterior oblique beams (beams 2 and 3 of Fig. 6-6) was rather critical, and that in addition to this the dose to the lung was still rather higher than desired. This arrangement (not shown) would likely, however, have been deemed acceptable had it not been easy to make more trials. It was next decided to eliminate beam 3 (Fig. 6-6) and add a wedge filter to beam 2. A number of weightings and angulations for this latter beam were tried, and finally beam 3 was again applied, but this time as a wedged field with the thick edge of the wedge down in the diagram of Fig. 6-6. A number of

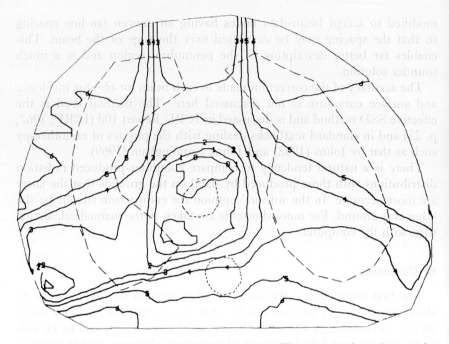

Fig. 6-9. An example radiation treatment plan. The target volume is near the centre of the chest. The lungs and the spinal cord are to be considered as sensitive structures.

adjustments in angulation and weight were made, and the distribution shown in Fig. 6-9 was chosen for the treatment. This diagram is a photograph of the incremental-plotter output. Only the body contour and internal structures have been touched up.

Beam number 1 is a 6 × 12 cm cobalt-60 beam directed downward at right angles to the anterior surface of the patient. The "given dose" or maximum dose from this beam (the term "weight" has been used above for this quantity) is 100 rads. Beam 2 is a 6 × 12 cm beam with a wedge filter inserted so that the thick end is uppermost in the diagram. It is at an angle of 110° to beam 1 and also has a weight (in this case the maximum dose on the center line of the beam) of 100. Beam 3 is again a 6 × 12 cm wedged beam, but with the thick end of the wedge down in the diagram. It has been given a weight of 65. The maximum dose within the target area is labeled 0 in Figure 6-9 and its magnitude is 114 rads. The curve labeled 1 corresponds to 110 rads, and so the area within this curve, which corresponds closely with the desired target region, receives between 110 and 114 rads. The curves labeled 2, 3, 4, and 5 correspond to 100, 80, 60, and 30 rads, respectively. The spine therefore receives everywhere less

6. Radiation Treatment Planning

than 80 rads (70% of the maximum tumor dose) and at the position of the cord less than 50% of the tumor dose. A portion of the lung on the left side of the diagram will receive 70–80% of the tumor dose, but this was deemed acceptable in view of the small extent of this high-dose region. There is also a small hot spot near the entrance of beam 2 where the dose is as high as the tumor dose. This, however, is not within a radiation-sensitive region of the patient and also is small in volume.

A second example application is shown in Fig. 6-10. Both diagrams

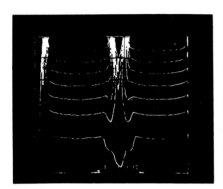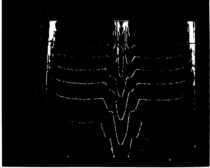

FIG. 6-10. Distributions where two beams are directed side by side on a plane surface.

show the oscilloscope display for two 10×10 cm cobalt-60 beams incident side by side on a plane surface. Occasionally, such an arrangement has to be applied to a patient, such as in the treatment of the central nervous sytem, where the equivalent of a very long narrow beam is required. The resultant distribution should be uniform over as much of the field as possible, and the gap between the beams, or region of overlap, becomes of critical importance.

In the photograph on the right the two beams are placed so that their edges at the surface are 1 cm apart. There is thus a gap at the surface, they match and produce uniform dosage at a depth of about 8 cm, and they overlap for depths greater than this. In the photograph on the left, the separation at the surface is 2.4 cm, producing a much larger low-dose gap and a uniform distribution only at a depth of almost 20 cm.

Neither of these distributions is ideal, and other spacings would have to be tried, possibly giving the beams a small inclination to the vertical as well, before the optimum arrangement would be achieved. Since manual calculation in the penumbral region is a very tedious and error-prone procedure, this example shows very well the advantages of computer over manual methods. However, it should not be supposed that the computer's usefulness is limited to unusual arrangements such as in these two examples. With summated distributions quickly and readily available, the majority of treatment plans might be given a closer scrutiny, possibly showing up hitherto undetected flaws.

It should be reiterated that radiation treatment planning is only a part of the overall treatment of a patient suffering from cancer. Nevertheless, it is expected that the optimization of this part of the care of the patient will lead to better results.

Acknowledgments

The Programmed Console project is the result of the efforts of a very large number of people, and the authors would like to acknowledge at least a few of them. The existence of the device owes much to the enthusiasm and midwifery of Dr. W. E. Powers of the Mallinckrodt Institute, St. Louis, and to the direction of Dr. Jerome R. Cox, Director of the Washington University Biomedical Computer Laboratory, St. Louis, where the Programmed Console was born. The "Enter Patient Contour" program and an assembler, which has been of inestimable help in writing the "Beam Generator" program, were written by M. D. MacDonald. The many evolutionary stages of the "Superimpose Beams" program were written by W. F. Holmes. A. M. Engebretson was mainly responsible for the two subprograms, "Plot Isodoses" and "Plot Plan." Elizabeth Van Patten wrote the "Enter Doses" program. The authors wish also to acknowledge communications with D. J. Wright of Temple University Hospital, Philadelphia, and Miss Marilyn Stovall of the University of Texas, M. D. Anderson Hospital, and Tumor Institute, Houston, other participants in the Programmed Console evaluation project. The encouragement of Dr. W. D. Rider of the Ontario Cancer Institute was much appreciated.

Programmed Consoles were provided to the Ontario Cancer Institute, Temple University Hospital, Philadelphia, the M. D. Anderson Hospital, Houston, and the University of Maryland, Baltimore, by the Washington University Computer Laboratories, St. Louis, from a grant from the Division of Research Facilities and Resources of the National Institutes of Health.

The authors gratefully acknowledge direct financial assistance from the Ontario Cancer Treatment and Research Foundation.

References

Clark, W. A., and C. E. Molnar. (1965). *In* "Computers in Biomedical Research" (R. W. Stacy and B. D. Waxman, eds.), Vol. II, pp. 35–66. Academic Press, New York.

Gupta, S. K., and J. R. Cunningham. (1966). Measurement of tissue–air ratios and scatter functions for large field sizes, for cobalt 60 gamma radiation. *Brit. J. Radiol.* **39**, 7–11.

IAEA (International Atomic Energy Agency) Vienna (1965). "Atlas of Radiation Dose Distributions," (E. W. Webster and K. S. Tsien, eds.) Vol. I. Single-Field Isodose Charts.

IAEA (International Atomic Energy Agency) Vienna. (1966). Computer Calculation of Dose Distributions in Radiotherapy, Report of a Panel. Technical Reports Series, No. 57.
IAEA (International Atomic Energy Agency) Vienna. (1967). "Atlas of Radiation Dose Distributions," (K. C. Tsien, J. R. Cunningham, D. J. Wright, D. E. A. Jones, and P. M. Pfalzner, eds.). Vol. III. Moving Field Isodose Charts.
ICRU (International Commission of Radiological Units and Measurements). (1962). ICRU Report 10d. National Bureau of Standards Handbook, **87**.
Johns, H. E. (1962). "The Physics of Radiology," 2nd ed. Charles C. Thomas, Springfield, Illinois.
Johns, H. E., and J. R. Cunningham. (1969). "The Physics of Radiology," 3rd ed. Charles C. Thomas, Springfield, Illinois.

IAEA (International Atomic Energy Agency) Vienna (1966), Computer Calculation of Dose Distributions in Radiotherapy, Report of a Panel, Technical Reports Series, No. 57.

IAEA (International Atomic Energy Agency) Vienna (1967), "Atlas of Radiation Dose Distributions," (G. C. Tsien, J. R. Cunningham, D. J. Wright, O. E. A. Jones, and B. M. Palmer, eds.), Vol. III, Moving Field Isodose Charts.

ICRU (International Commission of Radiological Units and Measurements), (1962), ICRU Report 10b, National Bureau of Standards Handbook, 87

Johns, H. E. (1962), "The Physics of Radiology," 2nd ed. Charles C. Thomas, Springfield, Illinois.

Johns, H. E., and J. R. Cunningham (1969), "The Practice of Radiology," 3rd ed. Charles C. Thomas, Springfield, Illinois.

CHAPTER 7

Some Data Transformations Useful in Electrocardiography[1]

JEROME R. COX, JR.

BIOMEDICAL COMPUTER LABORATORY, WASHINGTON UNIVERSITY, ST. LOUIS, MISSOURI

HARRY A. FOZZARD

DEPARTMENT OF MEDICINE, UNIVERSITY OF CHICAGO, CHICAGO, ILLINOIS

FLOYD M. NOLLE

BIOMEDICAL COMPUTER LABORATORY, WASHINGTON UNIVERSITY, ST. LOUIS, MISSOURI

G. CHARLES OLIVER

DEPARTMENT OF MEDICINE, WASHINGTON UNIVERSITY, ST. LOUIS, MISSOURI

I.	Introduction	182
II.	Transformations that Eliminate Redundancy	183
III.	Data Storage	186
IV.	A System of Processors for ECG Rhythms	188
	A. Ventricular Channel	190
	B. Atrial Channel	196
	C. Cycle Processor	197
	D. Program Details	198
V.	Results and Conclusions	200
	Appendix	204
	References	206

[1] This work was supported in part by the Division of Research Facilities and Resources and the Heart Institute of the National Institutes of Health, Bethesda, Maryland.

I. INTRODUCTION

In recent years it has become apparent through continuous monitoring of the electrocardiogram (ECG) that the majority of patients with acute myocardial infarction manifest cardiac arrhythmias, and that arrhythmias have been responsible for a large proportion of the fatal infarcts (Meltzer *et al.*, 1964). Where arrhythmias are detected, they frequently can be controlled by current medical methods, resulting in a considerable reduction in mortality (Day, 1963; Lown *et al.*, 1967). The treatment of patients suffering from these arrhythmias has been improved by establishing cardiac intensive-care units where continuous monitoring of the patient's ECG is possible.

The cardiac intensive-care unit has also made available objective means for evaluating methods of treatment of acute myocardial infarction (Gianelly *et al.*, 1967). With the establishment of these units, however, new problems have been created. A shortage of specially trained personnel has limited the effectiveness of new units. The lack of techniques for careful management and analysis of exceedingly large volumes of ECG records has hindered the initiation of systematic tests of many new drugs and therapeutic techniques.

To meet the needs emerging in cardiac intensive-care units, new monitoring and data-management techniques will be necessary. Present analog monitoring devices such as oscilloscope displays and rate meters with alarms will continue to be useful, but they have a number of shortcomings. Continuous observation of oscilloscope displays is impractical. Alarms designed to indicate high or low heart rate may sound inadvertently as a result of muscle noise when the patient turns or sits. Little help can be expected from these analog devices in the recognition of complicated arrhythmias or in the detailed analysis of ECG records needed for the study of new therapeutic techniques. Continuous analog tape records of the electrocardiograms of all patients in an acute cardiac care unit are so voluminous that they are difficult to store and process. Digital-computer processing of these ECG records is an attractive approach to these problems.

The cardiologist studies the ECG to gain an understanding of the underlying physiological events. Each cardiac cycle begins normally in a special pacemaker region, the sinus node. The signal is conducted to the atria and through the A–V node and conducting tissue to the ventricles. Arrhythmias may be the result of a variety of factors, such as the origin of a cycle in the atria, the A–V node, the conducting tissue, or the ventricles, or an alteration in their sequence or speed of conduction. Complex arrhythmias are difficult to identify in the surface ECG because waves corresponding to only some of the electrophysiological events appear there. The depolarizations of the atria and ventricles appear as the P wave and QRS complex, and the repolarization of the ventricles appears as the T wave.

A patient with an acute myocardial infarction may experience many sudden and dramatic changes in his ECG. These episodes can be brief, perhaps only a single beat in duration, or they may represent a stable but abnormal state of cardiac excitation. A digital-computer system designed to process ECG records should recognize these sudden changes, even though they may be rare and may be complicated by baseline fluctuations, minor components of muscle artifact, or irreducible electrode and electronics noise. Occasionally when the patient moves, muscle artifact may become so large that portions of the record are unrecognizable. These events should be identified for what they are, not mistaken for an arrhythmia.

The classification of rhythmic and morphological abnormalities directly from unprocessed ECG records leads to a signal space with unmanageable dimensionality. Some transformation of the ECG to eliminate redundancy is essential, but characteristics important in one context may be irrelevant in another. The human brain handles this context sensitivity in ECG analysis rather naturally, somewhat as it does in the understanding of language. In both these cases, however, context sensitivity produces serious added problems for computer processing.

An electrocardiogram is usually sampled with an analog-to-digital converter at a regular rate between 100 and 500 samples per second (sps). This data rate makes difficult an economical, continuous, and uninterrupted computer analysis of the ECG. Previous work has been done largely with finite records and analysis programs whose execution can take much longer than the duration of the record studied. Clearly, processing that keeps pace with the input-data rate is required for practical ECG monitoring systems.

In the paragraphs that follow, we describe briefly some of the techniques for processing ECG records that we have found useful in dealing with these problem areas. Some of these techniques are quite specific to our objective, ECG rhythm classification. Others may possibly be generalized to ECG morphology studies or even to work involving continuous processing of signals other than the ECG.

II. TRANSFORMATIONS THAT ELIMINATE REDUNDANCY

Not all of the information found in an ECG record is useful. How much information must be preserved and how much is redundant depends upon the ultimate purpose of the study. For that purpose a caricature retaining the essential features, but eliminating unimportant detail, will often be sufficient.

The generation of this caricature can be accomplished in a single processor or in a series or cascade of processors (Fig. 7-1). A comparison of these two processing schemes is analogous to the comparison of the production of an automobile by a single craftsman with its production by a series of workers on an assembly line. In the latter case each task is small, and a higher production rate is possible.

A simple analysis of the two processing schemes is revealing. In the single processor (Fig. 7-1a), the average time available to process one incoming bit is the reciprocal $1/R$ of the data rate R. For example, a data rate of 5000 bits per second (bps) allows an average processing time per bit of 200 microseconds (μsec), an insufficient amount of time when repeated scanning of data and complex logical analysis are required.

Consider now a cascade of processors (Fig. 7-1b), each producing a reduction in data rate of a factor greater than 2. It can be shown by a simple calculation that the time available to process one incoming bit is greater than $1/2R$ for each processor. Thus in this example each processor in the cascade, although implemented within a single digital computer, could be allowed at least 100 μsec of processing time per bit. Next consider the case in which each processor provides a factor-of-4 reduction in data rate. The processing time per bit can actually be doubled at each successive stage. For example, after ten stages 0.1024 second per bit would be available. Thus almost unlimited processing time per bit is available, provided the system includes a large enough number of processors, each with an adequate data reduction. With such a system continuous processing of the uninterrupted ECG is facilitated by the more efficient use of processing time and memory space.

The way a single digital computer shares processing time among the several processors is shown in Fig. 7-2. The first processor has the highest priority and is served whenever it has work to be done. Priorities decrease in the second and third processors. The last processor is served only when earlier processors are dormant. Because of its substantially lower input-data rate, interruptions to serve higher-priority processors cause no important delays. The space between the 45° line and the third processor line shows the margin of unused time available. Clearly, unused time still will be available when more processors are added, provided that an adequate data reduction is made by each processor.

An important consideration in the design of cascaded processors is a fidelity criterion. Such a criterion should be applied to the output of each processor to ensure that useful information not be lost. We have chosen

7. Transformations Useful in Electrocardiography

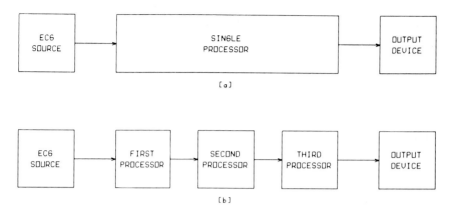

FIG. 7-1. System for the classification of ECG rhythms: (a) single processor; (b) cascade of processors.

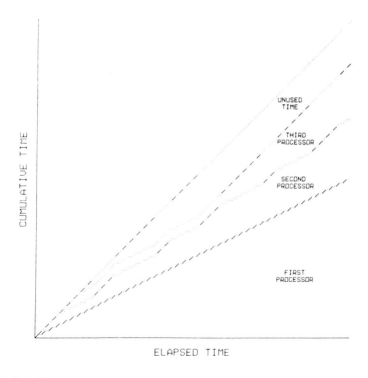

FIG. 7-2. Example of scheduling of time used by cascaded processors. Heavy lines indicate when each processor is in operation. Area just below 45° line indicates the margin of unused time.

to use the technique of reconstructing a caricature of the original ECG to test the fidelity of the processed data stream. For example, if an electrocardiographer can make as reliable a rhythm diagnosis based on data reconstructed from the output of a processor as from its input, we conclude that no important information has been lost in the processor. Clearly, questions about the ECG must be confined to the domain of interest specified when the processor was designed. To summarize, each processor in a system should be designed to avoid reproducing details that can be eliminated easily and are unnecessary to the ultimate analysis of the electrocardiogram.

The system design of a digital computer appropriate to carry out data transformations by means of cascaded processors is somewhat unconventional. Extensive data storage is not required because each processor need only retain very recent information. Only simple arithmetic is needed since most of the operations are addition, subtraction, scaling, or branching. The only input equipment required is an analog-to-digital converter. Output from the computer is a stream of data with a rather modest rate and perhaps an oscilloscope display.

Thus much of the equipment associated with conventional digital computers is unnecessary in this application. In fact a recent study indicates that the smallest computer capable of processing ECG signals from a single patient is more economical than the prorated cost of a more powerful computer shared among many patients (Cox, 1969).

III. DATA STORAGE

The storage of electrocardiographic data so that it may be retrieved quickly and easily is an important problem. An ECG that covers hours, days, or even weeks may be of interest to the clinician or research investigator. Storage of the entire ECG in its raw form is impractical for two reasons. First, the cost of the storage is large, but perhaps even more crucial is the difficulty of finding a wanted piece of data. If time of occurrence of the episode is known, simple time-code techniques allow retrieval. More often, however, both the clinician and researcher would like to see the electrocardiographic data arranged in a hierarchy. At the highest and most compact level the data are summarized so that only the gross state of the patient is presented. Below the highest level in the hierarchy the data may still be summarized, but with less abstraction and in greater detail. At lower levels in the hierarchy the data are presented in sufficient detail for reconstruction of the original waveforms according to some fidelity criterion. The lowest level in the hierarchy is the original ECG itself.

7. Transformations Useful in Electrocardiography 187

A clinician or researcher may wish to work downward from the most abstract level, selecting portions of the record of particular interest to him. Eventually, he might examine a reproduction of a section of the ECG or alternatively process the data to create a special summary of a set of measurements. Transformations that eliminate redundancy, yet allow reconstruction within a fidelity criterion, are extremely useful in conserving storage space and reducing the effort required to search and process such long records.

Here, as in the previous section, there is advantage to the scheme of cascaded processors shown in Fig. 7-1b. We have found it convenient to assign to each processor a circular buffer sized to hold all of the information necessary to that processor. The total memory space occupied by all of the circular buffers is much smaller than that required to store the original data for the epoch covered by the last processor. Each processor need review only the recent past. In fact, the storage requirement for a single processor (Fig. 7-1a) increases directly with the product TR of the epoch to be studied T and the data rate R. In contrast, a simple calculation shows that a cascade of processors (Fig. 7-1b) has a total storage requirement that increases logarithmically with TR. Thus, for large values of TR, enormous storage savings can be achieved with cascaded processors.

We think of the stream of data passing between processors as a stream of characters or symbols like that encountered in language processing. The context in which a character is found helps to determine the action of the following processor. As indicated earlier this sensitivity to context appears to be one of the important characteristics of the problem of processing electrocardiograms. The context broadens as the data stream passes from processor to processor. The data reduction achieved, however, by a processor makes available adequate context to the next processor without an unreasonable increase in storage requirements.

Figure 7-3 emphasizes the difference in storage requirements between the two processing schemes. For example, to identify an ECG rhythm may require a review of 20 seconds or more of the record. This corresponds to more than 100 kilobits of information from the source, a substantial storage requirement if a single processor is used. For continuous analysis double buffering will usually be required, and, if an attempt is made to eliminate edge effects, another doubling of the storage requirements may be necessary. Some arrhythmias may require the scanning of several minutes of data, an even more difficult storage problem. In the example of Fig. 7-3, the first processor in a cascade produces an output-data rate about one fifth the source rate. Successive processors each produce reductions in the data rate, so that the amount of information accumulated

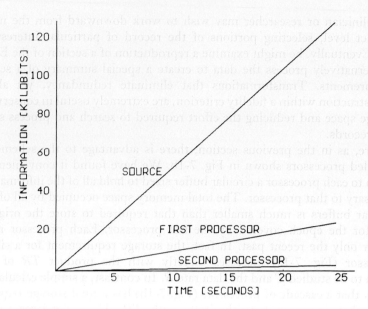

FIG. 7-3. Example of cumulative information produced by ECG source and by first two processors.

at the output of a third processor over a period of several minutes may be measured in the thousands rather than millions of bits. This amount of information is so much less than that produced by the source that it cannot be plotted clearly in Fig. 7-3.

New information always overlays the oldest information in a circular buffer; thus data are purged continuously as new data arrive. If experience shows that the circular buffer is occasionally not sufficiently large for its associated processor, the buffer size can be increased. Initially, most of our buffers have been larger than necessary. As experience has been gained, we have begun to shrink the sizes of those buffers having capacities that are not fully utilized.

IV. A SYSTEM OF PROCESSORS FOR ECG RHYTHMS

The general considerations reviewed in the preceding sections are based on our experience with cascaded processors in the handling of electrocardiographic data. To allow the reader to form a sharper image of the operation of such a set of processors, we describe below a portion of a system that has been designed for the on-line classification of electrocardiographic rhythms. Though the entire system is incomplete, the portion described seems to be useful in summarizing ECG data.

7. Transformations Useful in Electrocardiography

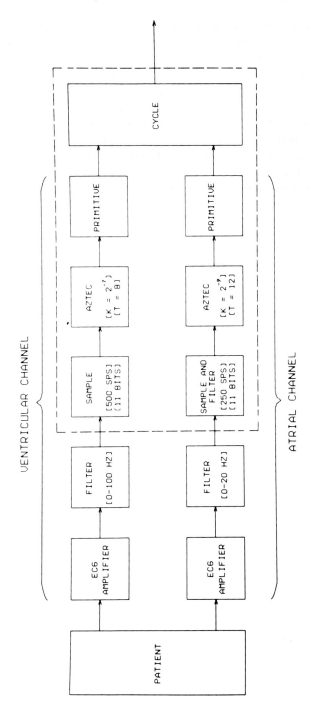

FIG. 7-4. Block diagram showing processing steps for the electrocardiogram. The upper channel is optimized for the detection of the ventricular (QRS) complex, the lower for the detection of the atrial (P) wave. The blocks enclosed within the dashed line are implemented by means of a digital-computer program. The AZTEC parameters K and T are defined elsewhere (Cox et al., 1968). The parameter K is referred to the full-scale range of the analog-to-digital converter and T is in milliseconds.

A block diagram of the system (Fig. 7-4) shows two parallel channels of cascaded processors. The upper channel is optimized to enhance the electrical sign of ventricular depolarization (the QRS complex). The lower channel is optimized to enhance the electrical sign of atrial depolarization (the P wave). We have chosen to make the two channels independent because for many abnormal rhythms the sequences of ventricular and atrial events are asynchronous. In particular we want to avoid a system that can detect P waves only when they are in a fixed temporal relation to the QRS complex. Figure 7-4 shows a separation in channels at the patient. There may, in fact, be separate and independent lead placements where one can be shown to enhance the P wave and the other to enhance the QRS complex. All of the data shown below have been gathered with a single pair of limb leads, usually lead II. Under these circumstances a single ECG amplifier is used with the two analysis channels connected in common to its output.

A. Ventricular Channel

An analog filter external to the digital computer passes signals in the range 0–100 Hz. We have ignored higher frequency components in the ECG because they seemed unimportant for rhythm analysis. Inside the digital computer, represented by the region inside the dotted line in Figure 7-4, the Sample Processor makes an analog-to-digital conversion at a rate of 500 sps. Lower sampling rates may produce samples that straddle sharp peaks and inflections. If these samples are used without careful signal reconstruction, distortion of the amplitude and onset of important waves such as the QRS complex can result.

An 11-bit analog-to-digital conversion is made, even though the dynamic range of a quiet ECG record would seem not to warrant this precision. Baseline fluctuations resulting from movement of the patient require that the limits of the conversion range be separated by about five times the QRS amplitude. We choose to limit the analog-to-digital conversion to 11 bits since this makes possible the calculation of differences of all possible samples without overflow in our 12-bit computer.

The Sample Processor produces at its output a stream of information with a data rate of 5500 bps. This rate is roughly comparable to the rate of the source shown in Fig. 7-2. This stream of information is fed to the next processor, AZTEC (Amplitude Zone Time Epoch Coding), which has the purpose of eliminating redundant data within long flat or sloping sections of the ECG record.

The first sample v_0 delivered by the Sample Processor to the AZTEC Processor sets initial conditions on two limits: $v_{max} = v_{min} = v_0$. (See

Appendix of Cox et al., 1968, for notation and a compact description of the processing algorithm.) Samples obtained at subsequent sampling instants are compared to these limits. If exceeded, a limit is replaced by the voltage just sampled. As long as the difference between the limits ($v_{max} - v_{min}$) does not exceed an experimentally determined "aperture" K, the fluctuating voltage is considered to be adequately represented by a constant voltage, or "line," midway between the limits. When finally a sample would necessitate separating the limits by more than the aperture K, the preceding average of the two limits is stored in the memory of the computer and called the value of the line. The time since the limits were initialized is stored as the duration of the line. In order to avoid long delays in the presentation of data to the next processor, a long line is terminated after 126 milliseconds (msec).

After each AZTEC line is formed and stored, the process is restarted by setting v_{max} and v_{min} equal to the latest sample voltage. When a signal of higher frequency and amplitude such as the QRS begins, the voltage samples will change rapidly, and lines of short duration will be formed. A series of lines, each 8 msec or less in duration, is considered to be adequately represented by a constant rate of voltage change, or "slope," as long as the difference between the values of adjacent lines does not change sign. The slope duration and the value difference between the lines bounding the slope are then saved.

The output of the AZTEC Processor can be thought of as a stream of symbols of two kinds: AZTEC slopes a_S and AZTEC bounds a_B. The latter are either AZTEC lines with duration greater than 8 msec or short lines encountered when a slope is terminated. The symbol a_B has two parameters associated with it, the value of the line and its duration. The symbol a_S also has two parameters associated with it, the value difference between the two lines that border the slope and the duration of the slope. Thus all AZTEC symbols carry along with them two parameters that may be used either for reconstruction of waveforms or in subsequent processors.

The purpose of the Primitive Processor is to examine a string of AZTEC symbols and recognize the ventricular events (QRS complexes). In Fig. 7-5 the alphabet of symbols $\{a_B, a_S\}$ produced by the AZTEC Processor is shown on the left and that produced by the Primitive Processor on the right $\{p_V, p_{\bar{V}}\}$. The symbol p_V identifies a ventricular event and carries with it parameters specifying the duration of the event and the shape of the complex in coded form. The symbol $p_{\bar{V}}$ identifies the interval between ventricular events and carries with it only the duration of this interval.

The Primitive Processor performs a fairly complex recognition task requiring an extended description. It is convenient to divide the Primitive

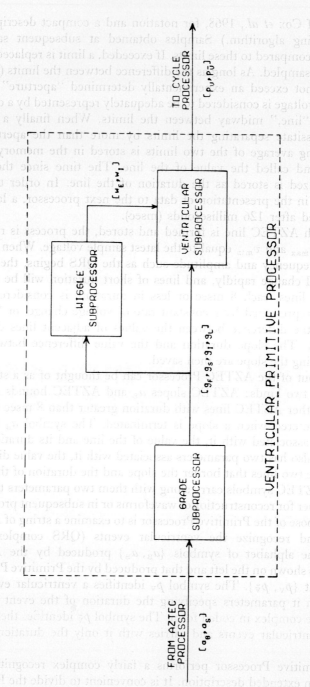

FIG. 7-5. Detailed block diagram of Primitive Processor for the ventricular channel. Input is a continuous stream of AZTEC symbols, either AZTEC bounds a_B or AZTEC slopes a_S. Output is a continuous stream of primitive symbols representing ventricular (QRS) complex p_V and portions of the electrocardiogram that are nonventricular waves $p\bar{v}$.

7. Transformations Useful in Electrocardiography

Processor into three subprocessors (see Fig. 7-5), each of which produces a stream of symbols from an appropriate alphabet. In a manner similar to the cases previously described, these symbols may carry along parameters that are useful later. The Grade and Wiggle Subprocessors carry out simple tasks and can be described easily. On the other hand, the Ventricular Subprocessor carries out a complex search that defies accurate description in prose. Thus a brief description of all of the subprocessors is presented here, while an accurate supplementary description of the Ventricular Subprocessor in a language invented for the purpose is deferred to the Appendix.

Prior to further analysis the noise content of 200-msec segments of an AZTEC string of symbols is evaluated. If at any time the input signal is off scale or if the average data rate for this segment exceeds 6000 bps or if the sum of the slope magnitudes exceeds a selected value, then the segment is labeled "chaotic," and no analysis is attempted. For nonchaotic segments, if the number of turning points is fewer than seven or if the data rate is less than 1500 bps, the segment is labeled "quiet," and a complete analysis is performed. If the segment is neither chaotic nor quiet, it is labeled "noisy," and only an abbreviated analysis is performed.

The purpose of the Grade Subprocessor is to identify flat sections of the record and to categorize slopes into three classes. All strings of contiguous AZTEC bounds a_B are joined together by the Grade Subprocessor into a single symbol g_F. A parameter carried along with g_F is the total duration of the AZTEC bounds that correspond to this symbol. The values of the AZTEC bounds that make up the g_F are discarded. AZTEC slopes fall into one of three grade categories, "small" g_S, "intermediate" g_I, or "large" g_L. The boundary value between intermediate and large grades is about one third of the amplitude of a typical QRS complex. The boundary value between small and intermediate grades is about one sixth the amplitude of a typical QRS complex. Carried with each of these three symbols g_S, g_I, and g_L is a parameter that indicates whether the direction of the grade is upward or downward. At times it is convenient to lump intermediate and large grades into the single category, "major" grade g_M.

The purpose of the Wiggle Subprocessor is to identify and group together symbols that represent vacillating portions of the record, portions that do not contain major grades nor long flat sections. This is accomplished by examining symbols from the Grade Subprocessors and identifying those strings that contain only small grades and flat sections with duration less than 100 msec. These vacillating strings, for an obvious reason, are called "wiggles." The class of wiggles that are bounded by major grades are called "inflections" or "extrema" according to whether the grades that bound them are similarly or oppositely directed. The output of the Wiggle Subprocessor is a stream of symbols having an alphabet that consists of an extremum w_E and an inflection w_I.

With the preliminary work done by the Grade and Wiggle Subprocessors, the Ventricular Subprocessor can begin the task of identifying QRS complexes. The onset and termination of the QRS complex must be recognized despite sudden changes in the shape of the complex and despite added complications due to baseline fluctuations and minor components of muscle, electrode, or electronic noise. Unusual QRS complexes provide a large share of the useful information to the clinician and investigator. Hours of accurate detection of normal QRS complexes may be less important than the identification of a few anomalous beats that possibly warn of more severe arrhythmias to come.

Figure 7-6 shows the AZTEC reconstruction of three consecutive QRS complexes that, despite their varying shape, are from a class of waveforms not infrequently observed in a cardiac intensive-care unit. The Ventricular Subprocessor begins its search for a QRS complex by scanning forward in its circular buffer looking for a large grade g_L. (The Appendix gives an exact description of the algorithm.) If one is found, its neighborhood is examined for an extremum w_E. First the search continues to the right, adding inflections w_I and similarly directed grades g_M, until the extremum w_E is found. An oppositely directed g_M to the right is then added, and the resulting string[2] is identified as an R wave p_R. If, for one reason or another, an R wave cannot be identified with a forward search, a backward search is then begun, examining the neighborhood to the left of the g_L. Inflections w_I and major grades g_M may be added until a w_E is encountered. A g_M to the left of the w_E completes the R wave p_R. The priority taken by the forward R-wave search helps to avoid identification of the trailing edge of a P wave and an oppositely directed leading edge of an R wave as the basic V shape associated with a QRS.

Now the basic V shape, here called p_R, associated with the QRS complex has been found. If the signal is quiet, the Ventricular Subprocessor proceeds to find portions of the waveform before and after the V shape that can be added legitimately to the QRS complex. The search for Q-wave elements proceeds backwards, adding up to two pairs of slopes and short flat sections. A similar search to the right of the basic V is carried out for S-wave elements. The entire "ventricular complex" p_V has now been

[2] The terminology used here for the components of the QRS is not entirely consistent with that used in electrocardiography. It is not necessary for the Ventricular Subprocessor to identify each of the classical components, providing the onset and termination of the entire QRS complex are recognized properly.

7. *Transformations Useful in Electrocardiography*

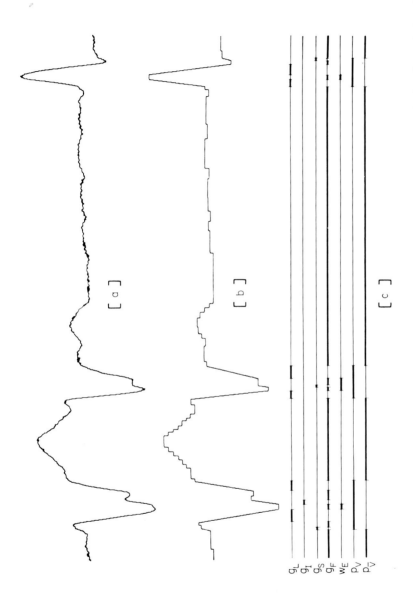

FIG. 7-6. Reconstructions of the output of the (a) Sample and (b) AZTEC Processors for the ventricular channel. Symbols produced in the Primitive Processor are indicated in (c).

assembled. For noisy signals the basic V shape alone without either Q-wave or S-wave elements is identified as the QRS complex p_V.

In all cases, the total duration must be less than 200 msec, and Q-wave elements must be preceded and S-wave elements must be followed by long flat sections. Those sections of the record that are not identified as a p_V are grouped together and are called "nonventricular waves" $p_{\bar{V}}$. For the ventricular channel and the sampled data of Fig. 7-7a, reconstructed input and output for the Primitive Processor are shown in Fig. 7-7b and c, respectively. The differences in the shapes of the QRS complexes are carried in coded form along with the primitive symbols p_V.

B. Atrial Channel

The problem of identifying the electrical signs of atrial depolarization (P waves) is, in many respects, similar to that of identifying the ventricular

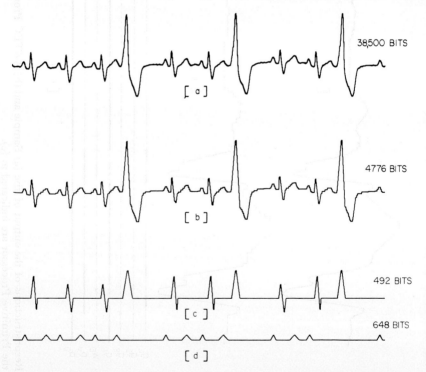

FIG. 7-7. For the ventricular channel (a) Sample, (b) AZTEC, and (c) Primitive Processor outputs, all reconstructed to resemble the original electrocardiogram. In (d) reconstructed output from the atrial Primitive Processor is shown. Given at the right is the total number of bits required to specify each trace. The total duration of each trace is 7 seconds.

complex. There are, however, some important differences. The P wave is considerably smaller in amplitude and therefore more subject to distortion by noise or other components of the ECG. Compensating for this difficulty, the shape of the P wave is generally less variable than that of the QRS complex, and the cardiologist is usually satisfied to measure its duration with less precision.

Consequently, a filter with a much lower cutoff frequency (20 Hz) is used. The Sample and Filter Processor within the digital computer for the atrial channel operates slightly differently than the Sample Processor in the ventricular channel. Samples are taken at a rate of 250 per second, and simple digital filtering is accomplished by averaging four successive samples before passing them to the AZTEC Processor.

The AZTEC Processor in the atrial channel operates in a manner quite similar to that in the ventricular channel; however, the parameters K specifying the AZTEC aperture and T specifying the minimum line duration are different. For the ventricular channel these parameters were chosen so that P and T waves would rarely produce slopes. The search for the QRS complex thus is simplified. In the atrial channel the aperture is one fourth that of the ventricular channel, and the minimum duration of a line is increased by half. These two changes cause P and T waves to be made up of sloping segments.

The Primitive Processor in the atrial channel also operates in a fashion similar to that in the ventricular channel. In its present form the major differences are the requirements that the P wave be upright and that no components be added to the basic V shape. Future plans include an extension of the algorithm to inverted and biphasic P waves. Reconstructed output from the Primitive Processor in the atrial channel is shown in Fig. 7-7d.

The Primitive Processor occasionally will mistake a T wave for a P wave. Careful examination of amplified ECG signals will show that these two waves at times can be remarkably similar in shape and duration. Thus there appears to be no error-free way to identify P waves that fall directly on top of T waves, since no added inflection points appear beyond those associated with T waves. Fortunately, P waves that appear slightly to the right of T waves can usually be detected by the Primitive Processor. Those that fall within the QRS complex are, of course, lost.

C. Cycle Processor

The input to the Cycle Processor is shown in Fig. 7-7c and d. The purpose of this processor is to merge the two streams of data, measure time intervals between atrial and ventricular events, and eliminate or minimize the effect of small fluctuations in the descriptive parameters carried within the stream of data. In addition, the Cycle Processor eliminates T waves mistaken for P waves by measuring their proximity to the QRS complex.

The operation of the Cycle Processor is the least well-defined of all the processors. In its present form its output has an alphabet of six symbols, three associated with ventricular events and three associated with atrial events. These symbols are the QQ interval c_{QQ}, the ventricular event itself c_V, and the interval measured to the preceding P wave c_{QP}. Two of these symbols, c_{QQ} and c_{QP}, carry parameters indicating the duration of these intervals. The symbol c_V carries a parameter describing the duration of the QRS complex and a parameter describing its shape. The symbols associated with the atrial event are the PP interval c_{PP}, the atrial event itself c_A, and the interval to the following QRS complex c_{PQ}. The parameters carried with these symbols are similar to those for the three ventricular symbols. In the case of the symbols c_{QP} and c_{PQ}, the interval may not exist; if so, an identifying code is carried along.

Data reduction is achieved in the Cycle Processor by grouping together all consecutive symbols of the same type in a manner similar to the line-forming section of AZTEC. For example, all consecutive c_{QQ} symbols form a line if the difference between the maximum and minimum intervals does not exceed an aperture. In this case the aperture is not constant but is dependent upon the previous value of the line. Reconstructed output from the Cycle Processor for four different rhythms is shown in Fig. 7-8.

D. Program Details

The transformations described above and indicated by the block diagram of Fig. 7-4 have been implemented in a small, stored-program digital computer. The word length is 12 bits, the memory size is 4096 words, and the memory-cycle time is approximately 3 μsec. The program necessary to carry out the activities of the Sample, AZTEC, Primitive, and Cycle Processors in both the ventricular and atrial channels occupies about half the memory. The remaining half is used for the various circular buffers. Experience has shown that in many cases the buffers can be reduced in size. We anticipate that about one quarter of memory could be made available for additional programs and data.[3]

---

[3] Recently a revised version of the Sample, AZTEC and Primitive Processor for both channels has been completed. This program along with the necessary circular data buffers occupies less than one-half the computer memory.

7. Transformations Useful in Electrocardiography

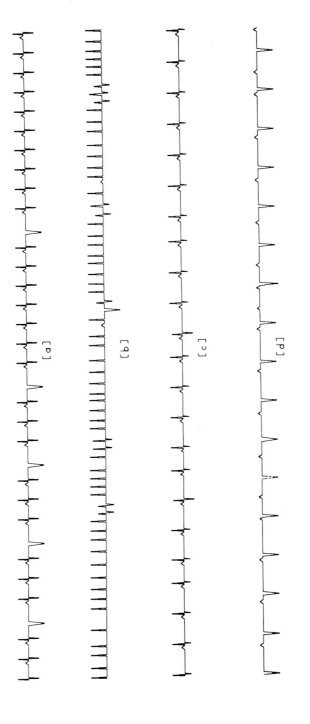

FIG. 7-8. Reconstructions of output from the Cycle Processor for examples of (a) normal sinus rhythm with premature ventricular contractions, (b) atrial fibrillation with aberrantly conducted ventricular beats, (c) normal sinus rhythm with premature atrial contractions, and (d) third-degree block with nodal rhythm.

No trouble has been experienced with the program failing to keep pace with incoming data. Ample spare time usually remains for the computer to create displays on an attached oscilloscope.

V. RESULTS AND CONCLUSIONS

The reconstructions of Sample, AZTEC, Primitive, and Cycle Processor data shown in Figs. 7-7 and 7-8 indicate that at no stage has information been lost that was crucial to a correct rhythm diagnosis. Examination of results from many ECG records disclose occasional errors in the processed data. The occurrence of these errors has been sufficiently infrequent to make their analysis quite difficult.

Data rates for the AZTEC, Primitive, and Cycle Processors are shown in Fig. 7-9 for about 60 seconds of typical ECG data. In all cases the rates are the total for the two channels. No satisfactory scheme has yet been achieved for eliminating the slight fluctuations that regularly occur in the coded description of the shape of the ventricular event. Thus the relatively small data reduction achieved by the Cycle Processor may be even less for some ECG signals. Overall, the system typically produces a data reduction of about 100 to 1.

Some of the results that can be obtained from the system are histograms derived from the Cycle Processor output. Several different histograms can be computed, and some illustrate the characteristics of particular rhythms in an interesting way. In Fig. 7-10 a two-dimensional histogram is shown with QRS duration plotted horizontally and QQ interval plotted in perspective. Figure 7-10a is the histogram for normal sinus rhythm interspersed with episodes of ventricular bigeminy. The premature beats associated with the bigeminy are at the lower right of Fig. 7-10a and show the expected reduced QQ interval and widened QRS. The beats following the compensatory pause are behind and to the left of the peak resulting from normal sinus rhythm. Figure 7-10b shows a similar plot for atrial fibrillation. Note the variability in the QQ interval and the stability of the QRS duration. The small secondary peak at the lower right is a set of aberrantly conducted beats.

Another method of plotting data from the Cycle Processor is shown in Fig. 7-11. The histograms are from a record of normal sinus rhythm with atrial premature contractions. The upper two histograms show that the QQ and PP intervals are unimodal with a tail to the left produced by the atrial premature contractions. The flatness of the tail indicates that no preferred coupling interval exists. The lower histogram shows the PQ interval, again unimodal and within normal limits. Figure 7-12 shows similar histograms computed for the ECG of a patient in third-degree

7. *Transformations Useful in Electrocardiography*   201

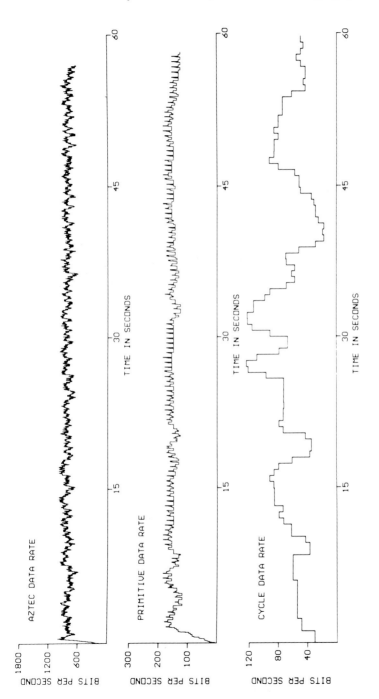

FIG. 7-9. Data rates produced by the AZTEC, Primitive, and Cycle Processors. The electrocardiogram for which these rates were obtained was an example of normal sinus rhythm with frequent premature ventricular contractions. In each case the combined data rate for the ventricular and atrial channels is shown.

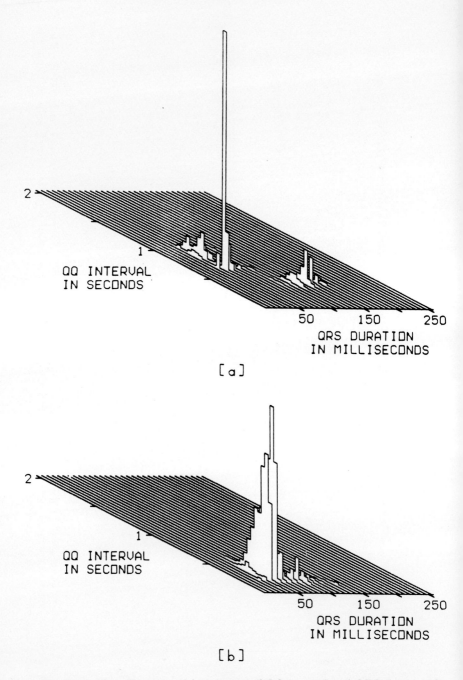

FIG. 7-10. Two-dimensional histograms of QQ interval and QRS duration for examples of (a) normal sinus rhythm with episodes of ventricular bigeminy and (b) atrial fibrillation with aberrantly conducted ventricular beats.

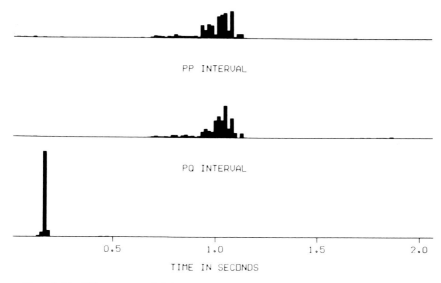

Fig. 7-11. Histograms of QQ, PP, and PQ intervals for an example of normal sinus rhythm with premature atrial contractions.

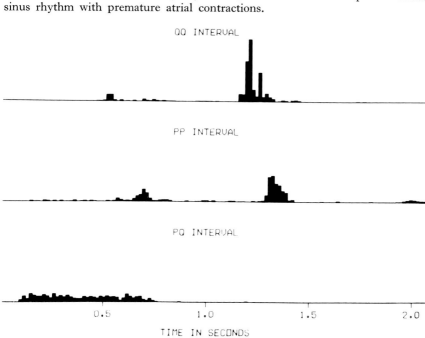

Fig. 7-12. Histograms of QQ, PP, and PQ intervals for an example of third-degree block with nodal rhythm. A few premature ventricular contractions can be seen in the histogram of QQ intervals.

block with nodal rhythm. Occasional premature ventricular contractions also can be seen in the histogram of QQ intervals. The histogram of PP intervals is multimodal, with peaks corresponding to integer multiples of the basic atrial interval. The second peak is the highest because many P waves were lost in the QRS complex or on top of the T wave. The PQ interval is essentially uniformly distributed up to a duration corresponding to the PP interval, thereby demonstrating the lack of coupling between atrial and ventricular events.

A scheme of cascaded processors used to eliminate redundancy in electrocardiographic signals has been developed with the aid of a small, stored-program computer. The reduced data can be reconstructed into a stylized electrocardiogram that is sufficient for rhythm diagnosis. We hope to use this output stream of information in a digital-computer monitor for a cardiac intensive-care unit. The data-reduction system can function continuously with rapid response. The reduced data are useful not only for monitoring purposes but also for evaluating the effects of new therapeutic measures.

APPENDIX

Algorithms for the processing of ECG signals can be completely described by extensions of some notations used in set theory. The extensions allow for some relations specific to strings of symbols (Ginsburg, 1966) and for a systematic presentation of the properties that define the set. In addition, recursive definitions such as those of Backus normal form (Backus, 1959) may be included. A complete and systematic description of the notation (Cox, 1969a) is not appropriate here, so just enough explanatory material is included to describe the Ventricular Subprocessor.

The input to the Ventricular Subprocessor is a sequence of symbols produced by the Grade Subprocessor and chosen from the alphabet $\{g_F, g_S, g_I, g_L\}$. A secondary input is a sequence of symbols produced by the Wiggle Subprocessor and chosen from the alphabet $\{w_E, w_I\}$. Portions of these sequences are scanned in search of a "forward R slope" $p_\rightarrow$.

$$P_\rightarrow = \{p : [p \in G_L] \vee [p \in P_\rightarrow W_I G_M]\} \qquad (1)$$

Here, the set $P_\rightarrow$ of all forward R slopes $p_\rightarrow$ is defined as the set of all strings p that cause the logical sum of the two bracketed constraints to be true. The logical sum is indicated by the symbol $\vee$. The string p thus, is a single "large grade" from the set G_L or a finite sequence of grades that belong to the set that is the concatenation of the set of forward R slopes $P_\rightarrow$ with the set of "inflections" W_I and the set of "major grades" G_M formed from the union of the sets of large grades G_L and "intermediate grades" G_I. Note that the presence of the defined set in the definition requires a recursive process to identify the set members. When a $p \in G_L$ is found, it is immediately true that $p \in P_\rightarrow$. The string to the right is scanned for a $w_I g_M$ and if one is found, a new $p \in P_\rightarrow$ is defined. The scan is then continued to the right for another $w_I g_M$. The recursion is continued until

7. Transformations Useful in Electrocardiography

no additional strings can be concatenated without violating the constraint implied by the second bracket in Eq. (1).

A "forward R wave" $p_{\overrightarrow{R}}$ belongs to the set

$$P_{\overrightarrow{R}} = \{p : [p \in P_{\rightarrow}W_E G_M] \wedge [\tau(p) \leq 200 \text{ msec}]\} \tag{2}$$

where $\tau(p)$ indicates the total duration of the interval covered by the string p. The logical product of the two constraints is indicated by the symbol $\wedge$. If no forward R wave is found satisfying Eq. (2), a search for a "backward R slope" $p_{\leftarrow}$ is begun.

$$P_{\leftarrow} = \{p : ([p \in G_L] \wedge [p \not\subset p_{\overrightarrow{R}}]) \vee [p \in G_M W_I P_{\leftarrow}]\} \tag{3}$$

The set $P_{\leftarrow}$ is defined with only slight differences from the set $P_{\rightarrow}$. The most important is the addition of the constraint $[p \not\subset p_{\overrightarrow{R}}]$; that is, p is not a substring of a forward R wave.

A "backward R wave" $p_{\overleftarrow{R}}$ belongs to the set

$$P_{\overleftarrow{R}} = \{p : [p \in G_M W_E P_{\leftarrow}] \wedge [\tau(p) \leq 200 \text{ msec}]\} \tag{4}$$

An "R wave" p_R is a member of the union of the sets of backward and forward R waves,

$$P_R = P_{\overleftarrow{R}} \cup P_{\overrightarrow{R}} \tag{5}$$

One or two "Q-wave" elements p_Q,

$$P_Q = \{p : [p = gg_F] \wedge [g \in G_S \cup G_M] \wedge [\tau(g_F) \leq 30 \text{ msec}]\} \tag{6}$$

may be added to the left of a p_R to get a "QR wave" p_{QR},

$$P_{QR} = \{p : ([p \subset g_F p] \wedge [\tau(g_F) > 30 \text{ msec}] \wedge [p \in P_Q P_R \cup P_Q P_Q P_R] \\ \wedge [\tau(p) \leq 200 \text{ msec}]) \vee ([p \in P_R] \wedge [p \not\subset p_{QR}])\} \tag{7}$$

Here the symbol $\subset$ may be read "is in the context" or "is a substring of." Similarly, one or two "S-wave elements" p_S,

$$P_S = \{p : [p = g_F g] \wedge [\tau(g_F) \leq 40 \text{ msec}] \wedge [g \in G_S \cup G_M]\} \tag{8}$$

may be added to the right of a p_{QR} to obtain a "ventricular complex" p_V.

$$P_V = \{p : ([p \subset pg_F] \wedge [p \in P_{QR}P_S \cup P_{QR}P_S P_S] \wedge [\tau(g_F) > 40 \text{ msec}] \\ \wedge [\tau(p) \leq 200 \text{ msec}]) \vee ([p \in P_{QR}] \wedge [p \not\subset p_V])\} \tag{9}$$

The portion of the ECG between p_V symbols, of course, becomes a "non ventricular wave" $p_{\bar{V}}$.

$$P_{\bar{V}} = \{p : [p \subset xpy] \wedge [x, y \in P_V] \wedge [p \notin P_V]\} \tag{10}$$

Thus the output of the Ventricular Subprocessor is a sequence of symbols from the alphabet $\{p_V, p_{\bar{V}}\}$.

References

Backus, J. W. (1959). *Proc. Intern. Conf. Inform. Process. Paris.* 125–132.

Cox, J. R. (1969). *In* "Future Goals of Engineering in Biology and Medicine" (J. F Dickson and J. H. U. Brown, eds.), pp. 196–204. Academic Press, New York.

Cox, J. R. (1969a). Internal memorandum, Biomedical Computer Laboratory, St. Louis, Missouri.

Cox, J. R., F. M. Nolle, H. A. Fozzard, and G. C. Oliver, Jr. (1968). *IEEE Trans. Bio-Med. Engr.* **15**, 128.

Day, H. W. (1963). *Dis. Chest* **44**, 423.

Gianelly, R., J. O. von der Groeben, A. P. Spivack, and D. C. Harrison. (1967). *New Engl. J. Med.* **277**, 1215.

Ginsburg, S. (1966). "The Mathematical Theory of Context-Free Languages." McGraw-Hill, New York.

Lown, B., A. M. Fakhro, W. B. Hood, Jr., and G. W. Thorn. (1967). *J. Am. Med. Assoc.* **199**, 156.

Meltzer, L. E., F. Palmon, M. Ferrigan, J. Pekores, H. Souer, and J. R. Kitchell. (1964.) *J. Am. Med. Assoc.* **187**, 986.

CHAPTER 8

Computation for Quantitative On-Line Measurements in an Intensive Care Ward[1]

JOHN J. OSBORN,[2] JAMES O. BEAUMONT,
JOHN C. A. RAISON, and ROBERT P. ABBOTT[3]

DEPARTMENT OF CARDIOVASCULAR RESEARCH AND RESEARCH DATA FACILITY,
INSTITUTE OF MEDICAL SCIENCES OF PACIFIC MEDICAL CENTER, SAN
FRANCISCO, CALIFORNIA

I.	Introduction	207
II.	Description of the Computer System	208
III.	Electrical Safety	216
IV.	Respiratory System and Analysis	217
V.	Cardiac Output	223
VI.	Calibrations	224
VII.	ECG Analysis	227
	A. Objectives	227
	B. Experience to Date	227
	C. Environmental Constraints	229
VIII.	Alarms	229
IX.	Exercise Laboratory	233
X.	Evaluation of the System	234
	References	237

I. INTRODUCTION

The amount of knowledge available in Medicine is rapidly outrunning the means of making all that knowledge useful in the care of patients. We will describe an intensive-care-ward measuring and monitoring system

---

[1] Supported in part by USPH Grant 5 PO1-HE 06311–07 and USPH grant 2 PO 7 FR 00241–03 COM and in part as a joint study project with IBM Corporation.
[2] Department of Cardiovascular Research.
[3] Research Data Facility.

that was developed to acquire, digest, and quickly present in useful form some important cardiac and respiratory data in very sick patients.

This system was designed with three major and quite separate aims, as follows:

(1) First, to accumulate new and useful data of a kind that is normally not available to doctor and nurse and that might contribute directly to the diagnosis and treatment of the patient. In other words, it was to extend the information available to the physician for his use in clinical care and in research.

(2) Second, to speed up the acquisition, processing, and analysis of presently accepted clinical data (blood pressure, temperature, and other commonly measured parameters) and present these data in collated form that could be more effectively appreciated.

(3) Third, to promote work simplification for nurse and doctor. It was planned to achieve this partly as described above by developing rapid methods of data compilation and partly by the provision of automated nurses' notes and the elimination of chores such as maintaining fluid-balance charts, etc., which might directly cut the manpower costs of illness. Further, the automation itself would have important effects in improving the quality of the data gathered. For instance, traditional methods of annotation by hand can be shown to be inaccurately time-related, and traditional notes are usually weakest during periods of crisis, when most needed. As planned, the system would make it possible to carry out sophisticated studies over long periods, without increasing the work load of nurses and doctors and with a precision superior to that of hand methods.

We also seriously considered allowing the system to execute therapeutic steps directly, such as replacing intravenous fluids lost; in other words, designing the system to "close the control loop." However, we made the tentative decision that, at least at this time, analysis *plus* control might be attended by instabilities that might harm the patient, so development of a control function was deferred. The experiences of other groups dealing with similar problems have been very helpful (Slubin and Weil, 1966; Damman et al., 1968; Lewis et al., 1966; Siegel and Del Guercio, 1967; Warner, 1965; Welkowitz, 1964).

II. DESCRIPTION OF THE COMPUTER SYSTEM

The physiological monitoring system consists of a number of different units in different locations, all connected to a central IBM 1800 computer and each doing basically the same types of data collection, computation,

8. Computation for Quantitative On-Line Measurements

and display. It has been partially described previously (Osborn et al., 1968). It includes four bed-study units for intensive-care patients, a mobile respiratory analysis cart for use in isolation rooms or to reach less accessible locations, a treadmill exercise-study unit with special display capabilities, and (under construction) a study unit for the catheterization laboratory. The functions and interrelations of these various units are shown in Fig. 8-1. In an intensive-care ward now being completed, all of the 14 bed positions have wiring installed capable of carrying any analog signals obtained to the computer for processing. Modular bedside input/output terminals for each bed are being constructed.

All have similar instrument capabilities, viz, ECG, three blood pressures, respiratory flow and pressure, respiratory P_{O_2} and P_{CO_2}, several temperatures, a closed-circuit television display for computer values, and a number of digital input switches and control buttons. The Eberhart arteriovenous oxygen analyzing cuvette (Eberhart, 1968) connects to the computer through one ward console, and a dye-dilution densitometer connects through any console.

The double unit in the intensive-care ward has been most extensively studied and used. This ward is a two-bed room, receiving mainly postoperative cases from open-heart surgery. The sensing and display system forms a cabinet of instrumentation that stands between the heads of the two beds. Wires to pressure transducers and electrocardiograph electrodes, and tubing to gas transducers pass over the head corner of the bed from the instrument complex to the patient, allowing access to both sides of the

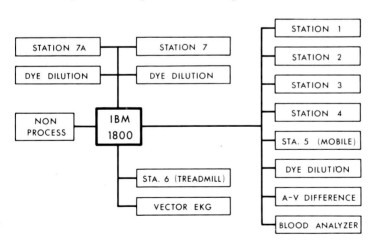

FIG. 8-1. Organization of sensing and computing units. A station consists of ECG, AP, VP, LAP, TEMP, RF, RP, P_{O_2}, P_{CO_2}, keyboard, and TV display. (See Table I for definitions of abbreviations.)

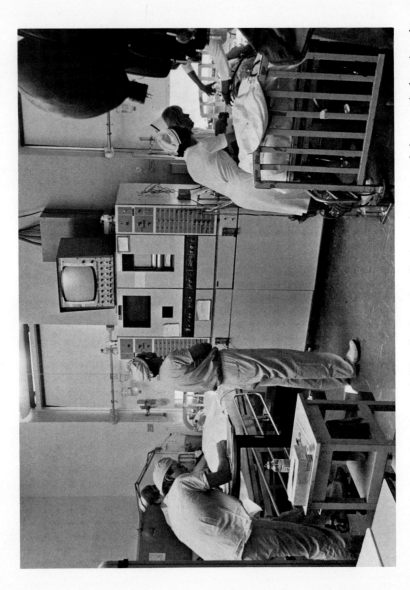

FIG. 8-2. The original two-bed study ward. The single analog display unit on top of the console has since been replaced by two separate units, one for each bed. The Computer system's display screen and the two-channel strip-chart recorder are built into the central console. Control switches are below and on both sides.

8. Computation for Quantitative On-Line Measurements

patient and free movement entirely around the bed, except for some restriction at the head. When a respirator is used, it is placed at the head of the bed on the side opposite to the instrument cabinet (Fig. 8-2).

The central console contains sensing and amplifying equipment and input and control panels for each patient, plus one analog oscilloscopic display for each patient and a single closed-circuit television screen for computed displays. A keyboard typewriter (IBM 1816) is in a recess in the corner of the ward and is used for commands to the computer, as well as for the generation of hard copy of nurses' notes, cardiovascular or respiratory computations, or other matters.

The two oscilloscopes (one for each patient) show arterial blood pressure, right and left atrial pressures (when available), and the electrocardiogram for each patient on a current basis. They are connected directly from the sensor amplifiers and so are independent of the computer. They thus fulfill the same purpose as the analog monitor displays commonly used in coronary-care wards.

The computer system's display screen is used to present a wide variety of information, ranging from current computed values of all measurements to temporal logs of past events and warnings of various sorts. The displays are presented on a 14-in. closed-circuit television screen from information generated on the faces of one of a series (three, being expanded to five) of Tektronix 564 oscilloscopes, each viewed by a TV camera. The information to be displayed is chosen as desired by manipulation of the switches or control buttons on the console or by typewriter entry. Some of the measured or computed parameters, the units in which they are expressed, and the identification abbreviations used are shown in Table I (p. 214). (Standard respiratory symbols are not compatible with the somewhat limited typewriter characters available).

Several typical displays are shown in Figs. 8-3, 8-4, and 8-5. The displays can be called or changed very rapidly or a series reviewed in sequence. The plots of variables against time can be shown for any number of hours from one to twenty four. In many of the standard numerical displays, several columns of data are presented showing changes over many hours. A heading on the screen always shows the appropriate bed number and the variables shown.

Typewriter entries are used for many kinds of information. For instance, it is important to know whether a ventilator in use is volume or pressure guaranteed and to know its gas-flow settings. A keyboard code has been devised for rapid input of such information with minimal learning by the operator. A similar type of code entry secures identification of a new patient to the computer system for logging purposes. Another permits the entry of blood-gas analysis results, if they are obtained from off-line instruments. In each entry, keyboard typing of a simple code, usually of

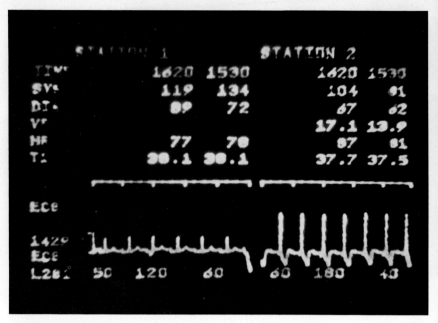

FIG. 8-3. Display of digital information on blood pressure, heart rate, and temperature for two beds, current and one hour earlier for each, plus current ECG for each.

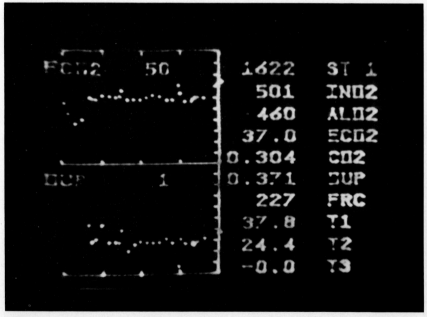

FIG. 8-4. Some current respiratory data, plus plot of expired P_{CO_2} and oxygen uptake over the last four hours for bed number 1. Note data and bed number at top of display.

8. Computation for Quantitative On-Line Measurements

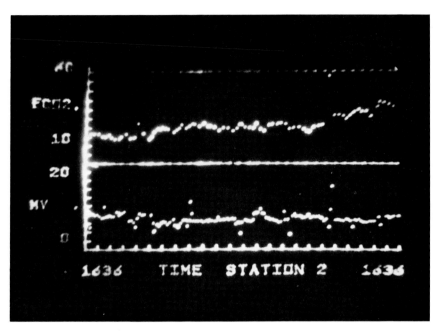

Fig. 8-5. Twenty-four-hour plot of expired P_{CO_2} and minute volume for patient in bed 2. Note slow development of respiratory acidosis. The physician might well now wish to call up the 24-hour plots of rate and volume to start to diagnose the difficulty.

two letters, is followed by a keyboard response listing in order and place the factors that should be entered—for instance, the time of actual sampling or the numerical values from gas analysis.

When blood-gas values are entered, the computer responds by searching the log for the nearest respiratory gas analysis on either side of the time of blood sampling, and it then types out the corresponding values of a variety of respiratory calculations including dead-space to tidal-volume ratio, ventilatory equivalents, and a number of other parameters. These values are also entered into the permanent log.

For permanent records, a 24-hour printout is made of all recorded data, and at the same time a graphic plot is derived using an IBM 1627 plotter, as illustrated in Figs. 8-6 and 8-7. These plots have been particularly useful in demonstrating unexpected trends. For instance, the plots of Fig. 8-7 (Osborn et al., 1968a) showed a malfunction of the respirator at about the 18th hour (near the right-hand side of the record), which resulted in severe hyperventilation, a dangerous drop in expired-carbon-dioxide concentration, a rise in heart rate, and a temporary deterioration in the patient's

TABLE I
Abbreviations and Units of Measurement for Measured and Computed Parameters

Abbreviation	Parameter	Units
SYS	Systolic arterial pressure	mm Hg
DIA	Diastolic arterial pressure	mm Hg
VP	Venous pressure	mm Hg
LAP	Left atrial pressure	mm Hg
HR	Heart rate (from ECG)	beats/min
PR	Pulse rate (from arterial pressure curve)	beats/min
T_1	Temperature probe no. 1 (others indicated by number)	degrees C
MV	Minute volume—ventilation	liters/min
TV	Tidal volume	cc/expiration
RRT	Respiratory rate	per min
CMP	"Total" compliance of ventilatory system	liters/cm H_2O
NER	Nonelastic airway resistance	cm H_2O/liter sec
WKIN	Work of inspiration during positive-pressure breathing	kg/min
OUP	Oxygen uptake	liters/min
RQ	Respiratory quotient	ratio
ECO_2	End-tidal P_{CO_2}	mm Hg
ALO_2	End-tidal P_{O_2}	mm Hg
INO_2	Inspired oxygen concentration	%
AP	Mean arterial pressure	mm Hg
APID	1st derivative, arterial pulse upstroke	mm Hg/msec
PRT	Pulse rate (from arterial pulse trace)	beats/min
HEM	Hemoglobin	g/100 ml
APO_2	Arterial oxygen tension	mm Hg
VPO_2	Venous oxygen content	mm Hg
FICK	Cardiac output by Fick method	Liters/min
RAV	Alveolar ventilation	liters/min
VD/VT	Dead-space–tidal-volume ratio	Ratio

condition, as noted by the nurse. Proper function of the respirator was restored and the respiratory values returned to normal.

The computer is an IBM 1800 with a 32,000-word 2-μsec memory, a 2310 disc storage unit with 3 discs of half a million words each, a 1442 card read punch, and a 1443 line printer. Also included are two 1816 keyboard printers, one 1053 printer, one 1627 plotter, 128 digital input positions (16 voltage sense and 112 contact sense), 32 process-interrupt positions, 192 digital output positions (160 "electronic contact operate" lines, 16 pulse output lines, and one 16-bit register output), 2 analog-to-digital converters, 2 solid-state multiplexers (32 inputs each), one low-level

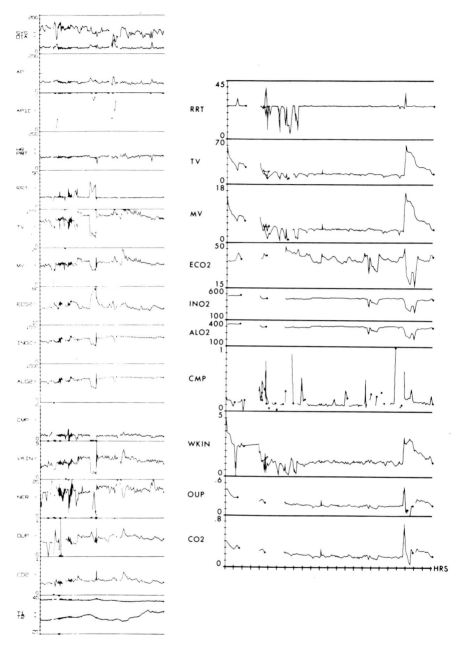

FIG. 8-6. One of the daily data plots. Detail is lost in this reproduction, but some impression can be gained of the wealth of information available.

FIG. 8-7. Part of a plot similar to that of Fig. 8-6 redrawn. Note the accidental hyperventilation beginning at about the eighteenth hour.

relay multiplexer (32 inputs), and six digital-to-analog–converter output lines.

Special hardware includes 3 display channels, each consisting of a storage oscilloscope and closed-circuit TV link, analog line-terminating amplifiers, and an external crystal oscillator for synchronizing analog inputs.

All programs operate under the control of the IBM 1800 TSX (time-sharing executive) system. At this writing the operating system is being converted from TSX to MPX (multiprogramming executive), which allows multiple core loads to be in execution simultaneously. Alarm analysis programs operate 5 times per second on data that have been obtained during the preceding 200 msec. Routine analysis programs and display-generating programs are executed automatically every ten minutes and additionally upon request. An interrupt heirarchy establishes priority among the several programs. When no process program is being executed nor in queue, control is turned over to nonprocess jobs such as report and plot generation, compiling and assembling new programs, or totally unrelated background-data processing. In normal operation about two thirds of the computer's time is available for such nonprocess jobs.

Every ten minutes the routine analysis programs calculate all the parameters shown in Table I (except those few requiring blood-gas measurements). When two patients are connected to one console, the respiratory measurements are carried out on them alternately. Under MPX complete analyses will be performed on up to 5 beds simultaneously. When more than 5 beds are on-line, requests will be queued for sequential analysis. Treadmill and catheterization laboratory will be analyzed every 30 sec, but other beds will be analyzed only every 10 min or upon demand.

III. ELECTRICAL SAFETY

Normal industrial safety practices are not adequate in an intensive-care environment in which patients may have indwelling saline-filled catheters and/or implanted pacemaker electrodes. To eliminate the hazard of inducing ventricular fibrillation in such patients, the monitoring equipment was required to meet a design objective such that in normal operation not more than 10 mV (with respect to room ground) would be developed anywhere that it could touch anyone or anything that could touch the patient. Under a single fault condition the voltage difference should not exceed 20 mV. To achieve this, all power in the recovery room is "floated" through an isolation transformer provided with a ground-current detector alarm. Thus if a power lead should come in contact with the frame of the monitor, it would merely ground that side of the power line and sound an alarm. Only a negligible current would flow through the safety ground connection, giving rise to only a few millivolts frame potential with respect to the room ground.

Special attention was given to modification of the pneumotachograph to meet this requirement, as well as to preventing overheating during no-flow conditions. The ECG leads are provided with series resistors and diode limiters, which not only protect the patient from a fault in the equipment but also protect the equipment during defibrillation or cardioversion.

The computer is located in another building, which differs from the recovery room in ground potential by 200 mV peak to peak. To avoid introducing the computer-room ground into the recovery room, all signals are transmitted double-ended to differential amplifiers at the computer multiplexer, which remove the common-mode ground potential while preserving ground isolation.

IV. RESPIRATORY SYSTEM AND ANALYSIS

A pneumotachograph is placed in the respiratory line. This may be at the endotracheal tube for an intubated patient on a respirator or in a face mask for a patient on a positive-pressure respirator or breathing spontaneously. This sensor has been modified to allow measurement of airway pressure as well as flow and to allow continuous sampling of the gas mixture passing through the pneumotachograph. The pneumotachograph is placed as close to the nasotracheal or tracheotomy tube as possible or sometimes in a tightly fitting face mask, and adds a total dead space of only about 30 cc. Respiratory flow, pressure, and gas composition are derived from the pneumotachograph sensor, gas analysis being carried out by rapid analyzers for CO_2 and O_2 (Elliott *et al.*, 1966). From these data oxygen uptake, carbon dioxide output, and respiratory quotient are calculated on a breath-by-breath basis, and various measurements of respiratory mechanics are carried out. The parameters of respiratory mechanics include compliance, nonelastic resistance, and work of inspiration.

Lung mechanics are usually computed only when a patient is on a positive-pressure respirator, from measurements of airway flow (pneumotachograph) and airway pressure. The compliance measured includes lungs plus thoracic cage. Such a measurement is useful because the compliance of the chest cage is relatively high (Deal *et al.*, 1968) and, in the absence of voluntary respiratory efforts, a severe drop in compliance is almost always due to changes in lung compliance or to the accumulation of fluid within the chest. In the same way "work of inspiration" calculated is the work as performed by the respirator. If the patient makes respiratory movements in phase with the respirator, the "work" reported will be lower, because as the patient assists the respirator, the respirator does less work. The computations can be carried out using intrapleural pressure (measured from a chest-drainage tube) or esophageal pressure, so that the compliances and resistances of chest wall and lungs can be measured separately, though we have done this only rarely.

The recording of lung mechanics has been particularly useful in helping the physician with the adjustment of respirators. The data can identify periods when the patient uses voluntary respiratory movements out of phase with the respirator, even when these may be barely apparent clinically. A convenient display has been the x–y plot of airway pressure (horizontal) against inspired volume (vertical). To minimize confusion, inspiration only is plotted, and a series of breaths can be presented by altering the zero baseline of pressure for each breath (Fig. 8-8).

The method of calculating oxygen uptake using breath-by-breath computation from respiratory flow and gas-concentration data was chosen because classical spirometry is impractical in patients who may be changing from one respirator to another or around whom the "clutter" of gas bags or large instruments may interfere with nursing care. The pneumotachograph provides a sensing system that is small and unobtrusive near the patient. The special problems of the use of pneumotachographs for long-term measurements (condensation, baseline shift) have been minimized by warming the pneumotachograph head electrically to a temperature of 39°C, by reverse flushing the differential-pressure and gas-sampling leads with warm air between analyses, and by rezeroing the differential strain gauge automatically before each 10 min analysis.

Figure 8-9 shows the arrangement used. All valves are under computer control. During analyses, valves V1 and V2 are closed. Valves V3 and V4 are set to connect the pneumotachograph to the pressure and flow gauges. V6 is closed and V5 is set to allow respiratory gas to be drawn into the analyzers at a rate of 1.5 liters/min.

Between analyses, V3 and V4 change over, causing the lines to the pneumotachograph to be flushed with dry air at a flow rate (adjusted by needle valve) of about 1 liter/hr. Also, V1 and V2 open, venting the gauges to room air and V7 is opened to air to avoid temperature change in the oxygen cell. V5 shifts to flush line C at 3 liters/min, and V6 opens to drain line D at 1.5 liters/min. Thus, both during analysis and during flushing, gas is withdrawn from the patient' airway at the same rate of 1.5 liters/min.

The calculations are performed as follows: upon start of analysis, no action is taken until the start of the next inspiratory half-cycle. This, as with subsequent half-cycle points, is tentatively identified when the flow signal changes sign, but to avoid confusion over flutter about zero, is not accepted until a specified minimum volume has been achieved without

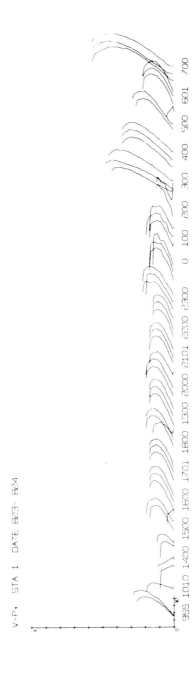

FIG. 8-8. Pressure–volume plots of inspiration only, over 24 hours. Volume is the vertical axis. Three breaths are chosen randomly every hour and plotted with a displacement of the pressure zero a fixed amount to the right each time. Note how the nonelastic or airway resistance increased around 2300–100 as shown by the pronounced bowing out of the half-loop. At 300 the patient is sucked out and placed on a different respirator. Airway resistance drops and tidal volume increases.

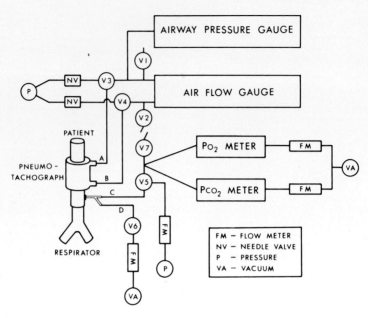

Fig. 8-9. Valving diagram for pneumotachograph blowback drying system.

subsequent flow reversal. Tidal volume is taken as the average of expiratory tidal volumes during the analysis period. Minute volume is taken as the sum of expiratory tidal volumes divided by the time spanned by the number of associated whole cycles. Work of inspiration is obtained by integrating the flow–pressure product over inspiration and dividing by the same time. As a first approximation, compliance is the average of each inspiratory tidal volume divided by its associated pressure difference. (See notes on dynamic compliance below.) The elastic work of inspiration is obtained by dividing the integral of the product of flow times the integral of flow, by the compliance.

The nonelastic work of inspiration is obtained by subtracting the elastic work of inspiration from the total work of inspiration. Since the nonelastic work also equals the nonelastic resistance times the integral of the square of flow, the nonelastic resistance may be found by dividing the previously obtained nonelastic work by the integral of the square of flow.

The measurement of compliance and resistance, as described above, can be refined. The computation described above derives compliance by dividing tidal volume by end-inspiratory pressure, which is taken as the pressure at the time of zero-crossing from inspiration to expiration. This gives "dynamic compliance," which is useful but which contains elements of nonelastic resistance and so may differ from the true steady-state

8. Computation for Quantitative On-Line Measurements

compliance (static compliance) by as much as 30–40% if the airway resistance is high. A closer approximation can be obtained when an Engstrom respirator is being used. This respirator is designed to maintain a constant inspiratory volume for the moment at the end of inspiration to allow for pressure equilibrium in the lung, so that a compliance calculated from the tidal volume and the airway pressure several milliseconds before the zero-crossing of flow gives a value that in most patients, is close to the true static value.

Another close approximation to true static compliance and true nonelastic resistance can be achieved by using a least-squares fit technique (Miller and Osborn, personal communication; Wald et al., 1967), with one of the classical lung-compliance equations to compute compliance and resistance. A study is now in progress to establish the limits of accuracy of this method.

The equation we have used is as follows:

$$P = \frac{1}{C}\int f\,dt + Rf + K$$

where P is the pressure (cm H_2O), C is the compliance (liters/cm H_2O), f is the flow (liters/sec), R is the nonelastic resistance (cm H_2O/liter sec), and K is the constant of integration (cm H_2O).

Pressure and flow signals are collected every 8 msec during inspiration. The volume at each point during the inspiration is found by integrating the flow up to that time. These parameters are then placed in two matrices representing the coefficients of the above equation and the least-squares fit is performed. Thus we solve for the lung compliance and nonelastic resistance simultaneously.

The calculation of respiratory gas quantities is complicated by the fact that although flow is sensed more or less instantaneously, approximately one-half second lag is involved before the corresponding gas sample can be pumped through approximately twelve feet of tubing to the gas analyzers. This lag is determined by the computer by comparing the start of inspiration with the drop toward zero of the CO_2 concentration. Flow values are then stored for this lag period before being multiplied by the corresponding oxygen and carbon dioxide concentrations.

A comment may be in order about the special technique of adjusting measured tidal volumes for the purpose of determining oxygen uptake. Stacy and Peters (1965) report expiratory volumes 10% greater than inspiratory volumes and automatically multiply inspired flow signals by 1.1 to obtain closed P–V loops. Osborn et al. (1968b) have examined errors due to temperature change (for a 10°C temperature rise the measured expired volume will exceed the inspired by 1.7% due to expansion and an additional 0.75% due to increased viscosity), gas composition (if air is inspired and 5% CO_2 expired, the expired volume will measure low 1.4% due to decreased viscosity), pressure (if inspiration is at 20 mm/Hg and expiration at atmospheric pressure, expiration will measure 2.6% greater due to expansion), and respiratory quotient (RQ). The cumulative effect of these is about 5 to 7% on tidal-volume measurement. However, a serious multiplication of error may occur in determining oxygen uptake at high concentrations. For example, if a patient inspires 60% O_2 and expires 57% O_2, a 1% error in tidal volume will cause a 20% error in oxygen uptake. For illustration, if tidal volume is 1000 cc, the oxygen uptake should be 30 cc per breath, the difference between 600 cc O_2 inspired and 570 cc O_2 expired. If, however, the expired volume is measured as 1010 cc, the computed expired O_2 will be 575.7 cc, giving an uptake of only 24.3 cc per breath. These errors are minimized by a two-step correction. The first step, which corrects for all errors, assuming unity RQ, consists merely of multiplying the inspired oxygen volume by the ratio of expired to inspired tidal volume. The second step is a reiteration using the corrected value. For example, consider first the effect of temperature alone. The measured inspired and expired volumes should be equal except for the volume-expansion and viscosity effects. Even if the air-stream temperature were measured, it would be difficult to make the proper correction because of uncertainty as to the heat-exchange effect of the pneumotachograph. However, the proper temperature correction would be that which just cancelled out the tidal-volume imbalance, which is precisely what the above multiplication does. The adjustment is made to be consistent with expiration so that final uptake figures will be expressed for body temperature. The same argument may be seen to correct for changes in viscosity due to changes in composition of the gas, provided the respiratory quotient is unity.

The pressure effect automatically cancels out, since the oxygen cell measures partial pressure of O_2 rather than percentage composition. Thus if the pressure were doubled, the flow measurement would be halved, but the partial pressure would also be doubled, giving the same indication of quantity of oxygen passing through the pneumotachograph.

Humidity is not a problem, since the respiratory gas is saturated at both inspiration and expiration.

In summary, the first step of this correction assumes that if the RQ were 1, inspired and expired volumes would be equal when adjusted for temperature, pressure, and concentration viscosity effects.

$$VIO_2^2 = VIO_2^c \times \frac{VE^m}{VI^m} \qquad (1)$$

8. Computation for Quantitative On-Line Measurements

In Eq. (1) and the subsequent calculations, the following notation is used: VI and VE are the inspired and expired volumes, respectively; O_2 and CO_2 represent oxygen and carbon dioxide, respectively; superscript m designates a value computed directly from a primary signal (initial measurement); superscript c designates a value first computed from an m value; superscripts 1, 2, and 3 indicate progressive calculations; and VO is the oxygen uptake.

As a second step, the oxygen uptake is recalculated on the basis of volumetric alterations produced by the first step:

$$VO_2^2 = VIO_2^2 - VEO_2^c \qquad (2)$$

(VO_2^1 is not used.)

Carbon dioxide production is calculated directly from the observed $VECO_2$, since no CO_2 is supplied during inspiration by the patient.

A new inspired volume, now adjusted for RQ, is calculated by subtracting oxygen uptake from, and adding CO_2 production to, the measured inspired volume:

$$VI^2 = VI^m - VO_2^2 + VECO_2^c \qquad (3)$$

The originally calculated inspired oxygen is now multiplied by the ratio of expired tidal volume to the new adjusted inspired oxygen volume:

$$VIO_2^3 = VIO_2^c - \frac{VE^m}{VI^2} \qquad (4)$$

A new oxygen uptake is calculated:

$$VO_2^3 = VIO_2^3 - VEO_2^c \qquad (5)$$

This procedure could be reiterated, but in practice a single series of calculations is adequate.

The correction also tends to correct for leaks, which are troublesome with face masks, although not usually with endotracheal tubes or tracheostomies. However, if the tidal-volume imbalance exceeds 10%, this fact is noted on the TV screen and in the log, indicating that the measurements are doubtful and the equipment should be corrected.

V. CARDIAC OUTPUT

Cardiac-output determinations are made both by the dye-dilution method and by using the Fick principle. A special oximeter developed in the Institute (Eberhart, 1968) measures light transmission at 660 mμ and 800 mμ wavelengths. The cuvette is mounted on a cart assembly, with a multiway rotary tap, withdrawal flushing pumps and reservoirs, and a recording amplifier system. It may be sterilized and attached directly to a patient's arterial and central venous cannulae or, as at present, so that the instrument may be available for more than one patient, syringes containing blood drawn anaerobically from these two sites may be attached simultaneously to the oximeter. The assembly has a small logic device that controls serial sampling, photometry, return of sample to the patient (if connected), and flushing of the cuvette. A single input/output cable is plugged into the bedside computer console. On pushbutton request for a Fick calculation, the following occurs: (1) A series of minute-by-minute oxygen uptake determinations is begun. (2) Two minutes later, arterial and venous samples are successively drawn through the oximeter cuvette. (3) The cuvette response analog output is sampled by the computer at programmed peak and trough points. (4) Oxygen consumption is measured following step (3) for a further 2 min. (5) Calculations are then made for mean value of 1-min oxygen uptake measurements, arterial and venous oxygen saturation, hemoglobin, A–V difference, and derived cardiac-output estimation. Output is immediate to the 1816 keyboard typewriter and to the display screen, and the data are entered in the patient log. This technique can be followed immediately by dye-dilution study using a commercial densitometer. At present, dye injection through the same central venous cannula is made by hand, a foot-operated relay simultaneously identifying the time for the computer. The analog output of the densitometer following automatic withdrawal from the same arterial cannula is sampled by the computer at low-level frequency. In a manner similar to that used by other computing dye-dilution systems, the computer extrapolates logarithmically the downslope of the dye curve from a point 70% of peak value and, after integration of the primary dye curve, makes and displays the following calculations: curve appearance and duration times, mean circulation time, blood volume between sampling points, and cardiac output by (a) the Stewart-Hamilton formula (Hamilton *et al.*, 1932) (b) the Dow formula (Dow, 1955)

$$\frac{\text{peak concentration (PC)} \times \text{PC time}}{3.0 - 0.9 \text{ PC time/appearance time}}$$

which has been claimed to be more reliable in low output states.

VI. CALIBRATIONS

The gas-analysis system is calibrated daily both directly and indirectly. The sensors for oxygen and carbon dioxide concentration are calibrated by drawing samples of room air and of known gas concentrations from measured gas mixtures through the sampling tube. The reference gases are measured by standard techniques (micro-Scholander for lower concentrations, paramagnetic oxygen analyzer for high oxygen concentrations). These analyses are also checked against mixtures prepared from pure gases by a proportioning pump (Godart).

Flow and volume computations are checked directly by drawing a known volume of air through the pneumotachograph sensing head with a hand-operated piston of known displacement, both with and without increased airway pressure, supplied by pumping into an elastic rubber bag. This gives calibration values for tidal volume, minute volume, compliance, and work.

Finally, the total computation of oxygen uptake and carbon dioxide output is checked daily using an "artificial lung," which consists of two equal syringes, each with a capacity of 600 ml equipped with a valve system so that one gas mixture (usually room air) is drawn in through the pneumotachograph sensing head and another gas (an "alveolar gas mixture" of known concentration) is "expired" through the same head from the other identical syringe (Fig. 8-10). Thus the pneumotachograph sensing head "sees" one gas during inspiration and another during expiration, and the volumes and concentrations of both gases are known. A measured volume is exchanged at several different respiratory rates, allowing the computed volumes to be compared with the known volumes. Over a series of 32 consecutive daily calibrations, the mean calculated oxygen uptake differed from the measured uptake by 1.4%, the standard deviation of the calculated uptakes was 5.7%, and the standard error of the mean calculated values was 1.0%.

As a further check on accuracy, occasional studies are made of oxygen uptake of patients on positive-pressure respirators, in whom a measurement series by computer is immediately followed by a direct spirometric measurement. In these studies the pattern of the respirator is identical for both series of measurements. The spirometer used is a "floating piston" recording spirometer, developed in our laboratories by Dr. Robert Eberhart, in which a low-friction piston sealed by a rolling diaphragm separates the two airtight sides of the spirometer. In use, the patient side of the spirometer is filled with oxygen, the other side is connected to the respirator, and the respiratory pressure is transmitted to the patient by movement of the diaphragm, whose recorded position also measures volume change. There is a CO_2 absorber on the patient side so absorption of oxygen on the patient side is shown as a change in volume just as with an ordinary basal metabolism unit. The advantages of the spirometer used are that it allows the measurement to be made while using the same respirator as the patient has been using, and its relatively small dead space makes only a minor change in the pattern of respiration supplied to the patient. An initial series of comparisons has been satisfactory.

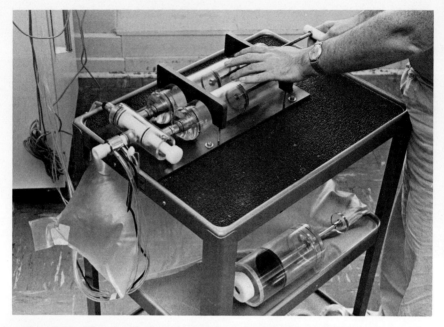

Fig. 8-10. The double-barrel syringe or "artificial lung." Room air or other inspiratory gas is drawn in through the pneumotachograph by one barrel, and an identical volume of "alveolar gas" from the bag under the syringe is expired from the other barrel. The smaller syringe used for calibration of volume and compliance measurement is lying on the lower shelf.

A review of our experience regarding automatic calibration may be useful. It was planned from the outset not to provide so-called "electrical" calibration such as switching a known resistor across a strain gauge. Present-day performance of electronic circuits provides sufficient gain stability to make this unnecessary, provided that occasional checks are made against known real pressures. However, there was provided an automatic zero calibration such that the computer could read and store as zero reference the output of any gauge when the computer was notified by pushbutton that the gauge was in fact open to atmosphere. This worked satisfactorily, but in practice did not provide sufficient improvement in operating convenience to justify the associated core requirements. There was also a small but not trivial instance of error due to inadvertent zero calibration by inexperienced personnel when the gauge was not open to atmospheric pressure. Consequently, we reverted to the system of having a technician daily calibrate by adjusting zero and gain to preset values.

A more elaborate automatic procedure was provided for the respiratory-gas calibration. When a "request calibration" button was pressed, the computer would activate valves to connect the gas analyzers first to air for 1 min and then to a standard gas mixture for 1 min., meanwhile calculating the average O_2 and CO_2 readings for each mixture. But this caused difficulties, mainly because of the required precise matching of four sampling lines (two beds, air, and standard gas). It proved impractical in practice to maintain this balance as new patient-sampling tubes were used, because of variation in the bore of the tubing. Consequently, we have reverted to a system in which the operator requests an air or a standard-gas calibration for each bed and manually ensures that the appropriate mixture is provided to the pneumotachograph until the TV display screen advises him that the calibration is complete and what values were obtained.

It is interesting to note that in both cases, the more sophisticated automated procedures initially implemented not only proved to be not worth the effort but were in fact slightly detrimental. While this may reflect merely the naivete of our initial approach, we suspect there may be a message here regarding simplicity versus oversophistication.

VII. ECG ANALYSIS

A. Objectives

Unlike computer programs that analyze ECG wave forms as an aid in the diagnostic process, the ECG-analysis programs developed for the system are designed to provide the physician with a real-time awareness that unusual and/or potentially dangerous events are occurring. Detailed analyses of the traditional kind are not really necessary for this purpose and are uneconomic in view of the environmental constraints described below. Changes in rate and development of ventricular ectopic[4] beats were designated as the information most useful in this context. Accordingly, the objective of the ECG-analysis programs is to provide for each patient monitored the following information: (1) heart rate; (2) ventricular ectopic rate.

B. Experience to Date

The program originally developed to satisfy the above objectives was based on relative changes in R-to-R interval. After nearly a year of experimentation it was judged to be a good indicator of heart rate, but its

---

[4] *Ectopic beat* is employed herein as meaning a beat caused by an impulse originating in the conductive system outside the node, or in cardiac muscle and occurring during the nonrefractory period of the cardiac muscle.

performance as an indicator of ectopic rate was unsatisfactory. No variation of parameters could be found that would satisfactorily filter out atrial arrhythmia based upon this criterion.

It then seemed essential to define another recognition scheme. Detailed examination of a multitude of strip-chart recordings made during the initial year of operation failed to identify *any* single measurable criterion that was either necessary or sufficient for ventricular ectopic identification by computer.

The presence of a ventricular ectopic is initially signaled to a human observer of an electrocardiogram by the fact that it "looks different." The absence of any one of the conditions usually present was found to be normally unconsciously overridden by the total visual impression imparted.

The following summarizes the results of examinations by cardiologists of strip-chart recordings that were identified as examples of ventricular ectopics:

(1) The voltage change from the peak of the R to the bottom of the S was often significantly greater in a ventricular beat than in a sinus one. However, it was frequently approximately equal and sometimes less.

(2) QRS duration was usually significantly longer, but in some patients this was not the case.

(3) Another variation was direction. A ventricular ectopic sometimes rose above the isoelectric line while a normal beat fell below, or vice versa.

(4) R–R interval usually was significantly shorter, but sometimes was not less than that expected.

In all cases observed (approximately 100 patients), *at least two* of the above phenomena were observed in each ventricular ectopic beat, but no two were observed in normal beats. Consequently, this last finding was experimentally employed as a ventricular-ectopic detector, any beat showing two or more of the four listed changes being tagged as ectopic.

While a description of the algorithm employed to detect QRS complexes is beyond the scope of the present work (McClung, personal communication), it is important to note that comparatives, rather than absolute values, are used throughout. Satisfaction of a criterion is determined by referencing averaged standards, which are frequently reestablished for each patient being monitored. The computer does not control gain; hence the percentage of full scale covered by the RS in different patients varies greatly. As to QRS duration, for some patients *all* QRS complexes are beyond "normal" limits, whether they are ventricular or atrial in origin. Moreover, there are patients whose "regular" beats produce such narrow complexes that ventricular ectopics, though longer, still fall within "normal" limits. An important advantage of the criteria explained above is that it makes unnecessary identification of any part of the waveform other than the QRS complex. This greatly simplifies the task and allows the program to operate within its environmental constraints. Note that no P-wave identification is necessary. The P wave is frequently unreliable, difficult to detect, or simply not present in the type of patient monitored by this system. P waves are important to the cardiologist when making a definitive diagnosis, but they are not what is looked for when scanning a CRT to identify premature ventricular contractions.

C. Environmental Constraints

(1) Human intervention and cooperation in controlling the program must be minimized. Any setup that gives a satisfactory ECG display on a cathode-ray tube is processable and should yield correct results. The program ignores lead and gain designations and needs only to determine when the leads are changed and/or when the gain is changed. This is accomplished by frequent reading of the lead and gain-control positions and by noting any setting changes. The program will reestablish the standard heart beat in the 1.2 sec following the change. Every 15 sec thereafter, an average is taken to establish a new standard. It is not necessary for medical personnel to provide information about waveshape, amplitude, or rhythm.

(2) Continuous and simultaneous monitoring of several beds must be provided at minimal cost in system resources (computer time and memory space), as several tasks are concurrently performing analysis of other signals. Accordingly, the program processes 200 msec of data at a time without transcription to external (disc) storage. The amount of time required to process these data ranges from 4 to 14 msec per bed, depending upon the presence or absence of a QRS in the input.

(3) There is no analog or digital preprocessing of data before they are received by the program. It operates successfully in the presence of 60-Hz noise from electric blankets, respirators, etc. Baseline shifts do not affect reliability because only the lengths (voltage change) of line segments in the QRS are used. "Where they are" with respect to the full scale is irrelevant. Artifacts are difficult to recognize, but reasonableness tests on a tentative QRS complex and RR interval help to eliminate faulty analysis.

VIII. ALARMS

An important function of an on-line monitoring system must be to provide alarms when there are threatening changes in the parameters being measured.

The first alarms provided in the system were similar to limit alarms on commercial analog monitor systems (high and low ECG and arterial pressure values) which activated light and sound signals, together with a written explanation of the limit exceeded in the ICU display CRT. Both physiological channels provided the commonly experienced and unacceptable excess of false alarms from artifacts. A detailed study confirmed that *all* false alarms arose from extrinsic causes, patient or operator originated and not within the sensing or data-processing components of the system (Raison *et al.*, 1968). Since this form of alarm is traditionally concerned with alerting staff to a cardiovascular crisis, it was possible to make use of the data-correlation function in the system to require the coincidence of exceeding alarm-limit settings in both ECG and vascular channels, which effectively reduced false alarms from 330 to 8 in a similar period of study. These 8 were caused by inadequately set limits and could be ignored. During the following year there were no failures due to nonoperation in critical situations. Limits could be changed by keyboard entry; the set values were shown at the bottom of every CRT display, with identification of those exceeded if an alarm situation arose.

Such a program is, however, extremely demanding of computer time. Without analog signal preprocessing, it requires continual analysis (in consecutive segments of 15 sec) of both arterial and ECG signals in this system, at high sampling rates, with a comparison procedure also continuously operating. This takes up to 20% of available computer time, observing only two beds, which might perhaps be considered in terms of $2000 monthly computer time rental. In a research study ward and during prototype development of a computer system, it is still considered necessary at this time to have a trained nurse for each acutely ill patient. In practice the few such critical cardiovascular emergencies have always been noted by the bedside nurse from traditional analog displays as early as the computer alarm operation. We have therefore, for the moment, abandoned the operation of this alarm system. Whether this would be a justifiable omission in any nonresearch scheme or whether it would be acceptable to other medical staff is debatable, but the economic implication should not be overlooked, compared with the cost of hiring additional trained observers. Use of signal preprocessing should greatly reduce cost, however.

Alarms, as so far described, can be considered the first grade of a series, progressively more complex, of diagnostic analyses. The second group of "alarms" are those occasioned by a change beyond alterable limits or of preset degree, usually in a single physiological parameter, either measured or calculated. They are of occurrences sensed by the system that should be made available for clinical consideration or that frustrate its own care-support ability or research accuracy. Output is by a text CRT display, and

8. Computation for Quantitative On-Line Measurements 231

several appear in the hard-copy log in order to qualify the recorded data values, if they may be used in research. We report limit infringements of heart rate or mean arterial pressure in this category of alarm. "Arterial line damped," signifying damping of the arterial waveform, is displayed if pulse pressure falls to less than 20% of mean arterial pressure, calling usually for flushing of the intravascular cannula. Other systems use relationship of pulse pressure to systolic pressure, approximation of systolic and diastolic pressures within absolute limits, mean arterial upstroke first derivative or rise and fall waveform time ratios, but without evidence that these are more reliable as an index of damping.

The display "TV unbalanced" is shown if inspiratory and expiratory tidal volumes measured by the pneumotachograph differ by more than 10%. Such a difference calls for a review of airway connections for leaks and rejection of current respiratory data for research, although trend significance is often retained by such values as respiratory gas exchange when the tidal volume error observation is only slightly greater than the set limit. The following will be added: a fall in total chest compliance below 0.020 l/cm H_2O, a rise of maximum inspiratory pressure beyond preset value during mechanical ventilation (denotes airway obstruction), and the reporting of the patient's being out of phase with the mechanical ventilator (work of inspiration greatly increased with erratic changes in total compliance). This category is valuable for temporary use, in order to build up an experience of the significance of any particular degree of signal variance in relation to some noted or anticipated clinical change, with a view to incorporation of that alarm as a routine function or as part of a more complicated diagnostic program.

We have by no means completed a review of the significance of changes in all the data already available to us and share the common experience of being bemused at times by huge arrays of numbers whose interrelationships are not immediately clear. And yet it remains a prime objective of the system that specific identification and naming of pathological processes shall be achieved faster and/or more precisely than by traditional methods, rather than the mere presentation of masses of figures. While the general terms of derangement in such conditions are not difficult to describe, there is considerable difficulty in describing them precisely enough and in the correct order for fully automated diagnosis. We must build up a program of probable diagnoses containing its own editing features.

Thus the pathway illustrated in Fig. 8-11, producing possible diagnoses of low effective circulating blood volume ("hypovolemia, give blood") cardiac tamponade, or primary cardiac failure (with a "score value" suggesting a greater likelihood of tamponade) is organized to operate in real time. It simultaneously compiles a printed log categorizing the specific

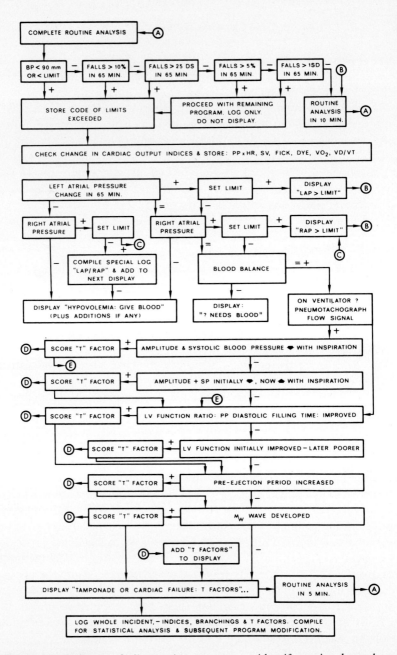

FIG. 8-11. Pathway of diagnostic program to identify major hemodynamic disturbances. The sign + means "has increased" or "yes," according to the context; " = " means "is unchanged"; " − " means "has diminished" or "no," according to the context. At each step, comparison is made with data collected over the last 65 minutes, which value may be altered; similarly, the values expressed within decision slots may be altered.

data points that activated the diagnosis. It should then be possible to correlate this with clinical findings and eliminate pathways or unsuitable limits that have given rise to "false alarm" diagnoses.

Before routine clinical application, one further step is desirable for a period. With a displayed diagnosis should appear a request for confirmation or denial (together with an "event registration number") by clinical staff through the terminal, immediately or later. "Limits" are values that can be set into the system for an individual patient's observation and care via keyboard or terminal.

The program compiles a log of its own operation, in which the route at each branching is coded. It similarly codes any or all of the indices of cardiac output that have changed and logs "T factors" (those that weight a diagnosis in favor of tamponade rather than primary cardiac failure). Every diagnostic program event will therefore acquire a set of code numbers of its pathway. From these the computer can compile, cumulatively, a statistical analysis of valid and invalid parts of the program before clinical application.

IX. EXERCISE LABORATORY

The exercise laboratory is a special development of the central computer monitoring system (Elliott *et al.*, personal communication) and shares the computer and many of the programs. This laboratory consists of a treadmill capable of various inclinations and speeds and electronic equipment to measure rate of flow of air, percent of oxygen expired, percent of carbon dioxide expired and ECG's. The sampling rate is the same as it is in the patient-monitoring system, and the same algorithms are used to detect beginning and ending of an inspiration for the purpose of calculating oxygen uptake, carbon dioxide production, respiratory rate, etc. The analysis of the ECG's is also by the same ECG algorithm as in the general system.

Before a study of a patient takes place, a technician calibrates the flow, oxygen, and carbon dioxide signals. The patient is then fitted with ECG leads in a manner that will allow him mobility. A headpiece is attached to the patient which holds a pneumotachograph with mouthpiece and connecting tubes, and the patient is seated in a chair on the stationary treadmill. At this point, the technician starts the study by turning a selector switch to the proper position and pressing a button, which the computer senses. The patient is then monitored during a prescribed resting period in the chair, during an exercise period with the chair removed, and during another resting period in the chair. A medical doctor is present during the actual study.

During the study there is a continual sampling of air flow, percent of inspired oxygen, percent of expired carbon dioxide, and ECG signals. At approximately 30-sec intervals the results of this information are analyzed and displayed on the CRT next to the treadmill. From the basic quantities measured the following parameters are calculated and displayed: oxygen uptake, carbon dioxide production, respiratory quotient, respiratory rate, heart rate, tidal volume, minute volume, end expired carbon dioxide partial pressure, and oxygen pulse. Also, an average (centered on the R wave) of ten seconds of ECG's is displayed with the other parameters. In addition to the display during the study, this same information is stored on disc memory by the computer and is available for processing. On a regularly scheduled basis the next morning, all of the parameters from the study, including the time, are printed. There is also a plot of each parameter versus time and each averaged ECG for the entire study (Fig. 8-12).

This information could be printed and plotted immediately after the study. We have found it adequate to deliver the information to the interested physician the next morning.

X. EVALUATION OF THE SYSTEM

In one of its primary functions (to provide new data for research purposes) the system has been very successful. New and interesting findings in the "natural history" of acute cardiac disease have developed and are being assembled for separate publication.

Its second function, as a clinical monitoring and early warning device, has been highly successful in some ways and needs further modification in others. On the positive side there have been many examples of clinical understanding due to the computer system where the diagnosis would probably never have been known without it. Particularly, the incidence of "unexplained" cardiac arrest has diminished almost to zero during its use. Two short examples will show why.

In a 73-year-old man, sudden cardiac standstill without clinical warning signs was clearly shown (by study of the 1627 plot of variables) to have begun four hours earlier, when an improperly set respirator led to respiratory acidosis, voluntary respiratory movement with increased oxygen uptake, gradually increasing oxygen debt, and finally the "unexplained" standstill.

As a second example, in a patient with cerebral edema, the decision was made to use whole-body hypothermia to reduce metabolic demand. When the cold blanket was turned on, the oxygen uptake rose from approximately 290 cc per minute to 650 cc per minute and his condition deteriorated

8. Computation for Quantitative On-Line Measurements

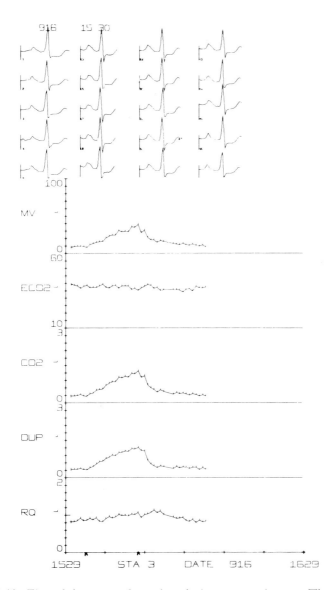

Fig. 8-12. Five of the many plots taken during an exercise test. Those shown include minute volume, expired P_{CO_2}, volumetric CO_2 output, volumetric oxygen uptake, and respiratory quotient. Above them are a few of the averaged ECG complexes computed and displayed every 30 sec. Time marks are given every 10 min. Start and stop of the exercise period is marked by blips on the time baseline.

precipitously. The patient had received sedation; there was no obvious shivering. Close inspection, however, after the increased oxygen uptake was observed, showed almost imperceptible fibrillatory movements of skeletal muscles, which might be called "subliminal shivering." With a much increased dosage of sedation, the oxygen uptake dropped back to below control levels.

In other patients, hemothorax has been diagnosed by the occurrence of a sudden fall in chest compliance before it was evident clinically. And finally, the "trend plots" on the display screen have certainly alerted the attending personnel to changes of one kind or another in a large number of patients, with results that are hard to document firmly, but which were certainly to the patient's benefit.

On the negative side, however, the very multiplicity of data available has caused problems in interpretation. We have discussed some of these in the section on alarms. It has become evident that important changes in one or more variables are often lost in the mass of information available. To make full clinical use of the system it will be necessary to implement a large number of what might be called "diagnostic programs," in which the computer senses changes in single variables or in patterns of variables and responds with appropriate diagnoses written in real language, identifying the specific changes as well as their possible meaning. A start in this direction has been described under "Alarms," but this type of function must be much expanded.

The impact on nursing staff has, in general, been favorable. Full impact on nursing cannot be evaluated until further progress is made in automating the information usually contained in nursing notes and, in general, reducing the work load of the nursing staff.

A final philosophical note: We try to think of the system not as a mechanical doctor or nurse competing with us, but as an extension of our senses and our memories. Thus it is not designed to take over our functions, but rather to provide us the means to function more intelligently for the benefit of the patient who, after all, pays the final bill.

Acknowledgments

This chapter was written by only four authors, but the system described was created by a larger group of workers, many of whom contributed important parts of its conception or construction and whose work should be recognized. We cannot list all contributors, but particularly need to mention Dr. Jerome A. G. Russell for his enthusiastic and penetrating inspiration and leadership while he was Director of the PMC Research Data Center; Mr. Stanley Elliott, who built the analog prototype from which the respiratory system was developed; Dr. Frank Gerbode for constructive support; Mr. William Radke for patient engineering; Mr. Bill Gilmore, Mrs. Dianne McClung, and Mr. Charles Miller for thousands

8. Computation for Quantitative On-Line Measurements

of hours of imaginative programming; Dr. Franz J. Segger for important development work; Drs. Donald Hill, Robert Popper, Alvin Hackel, and William Armstrong; Nurses Miss Paula Naylor, Mrs. Kathren Martz, Miss Sandra Geiger, and many others who know their own contributions.

References

Dammann, J. F., D. J. Wright, and O. L. Updike, Jr. (1968). Physiological monitoring: its role and contribution to patient care. *Natl. Instr. Soc. Meeting, 6th Pittsburgh, Pennsylvania.*

Deal, C., J. J. Osborn, E. Ellis and F. Gerbode. (1968). Chest wall compliance. *Ann. Surg.* **167**, 73.

Dow, P. (1955). Dimensional relationships in dye dilution curves from humans and dogs, with an empirical formula for certain troublesome curves. *J. Appl. Physiol.* **7**, 399.

Eberhart, R. C. (1968). An automated sampling whole blood photometer. *Biomed. Sci. Instrum.* **4**, 197.

Elliott, S. E., W. Armstrong, and J. J. Osborn. Personal communication.

Elliott, S. E., F. J. Segger, and J. J. Osborn. (1966). A modified oxygen gauge for the rapid measurements of PO_2 in respiratory cases. *J. Appl. Physiol.* **21**, 1672.

Hamilton, W. F., J. W. Moore, J. N. Kinsman, and R. G. Spurling. (1932). Studies on the circulation. *Am. J. Physiol.* **99**, 534.

Lewis, F. J., R. Shimizu, A. L. Scofield, and P. S. Rosi. (1966). Analysis of respiration by an on line digital computer system: clinical data following thoracoabdominal surgery. *Ann. Surg.* **164**, 547.

McClung, D. Personal communication.

Miller, C. W., and J. J. Osborn. Personal communication.

Osborn, J. J., J. O. Beaumont, J. C. A. Raison, J. A. G. Russell, and F. Gerbode. (1968a). Measurement and monitoring of acutely ill patients by digital computer. In press.

Osborn, J. J., F. J. Segger, S. E. Elliott, and F. Gerbode. (1968b). Oxygen uptake and pulmonary function in critically ill patients by analog computation. In press. *Med. Res. Eng.*

Raison, J. C. A., J. O. Beaumont, J. A. G. Russell, J. J. Osborn, and F. Gerbode. (1968). Alarms in an Intensive Care Unit: An Interim Compromise. *Computers Biomed. Res.* **1**, 556.

Siegel, J. H., and L. Del Guercio. (1967). The diagnosis and therapy of shock using physiologic correlates quantified with a bedside computer. *Bull. N.Y. Acad. Med.* **43**, 424.

Slubin, H., and M. H. Weil. (1966). Efficient monitoring with a digital computer of cardiovascular function in seriously ill patients. *Ann. Internal Med.* **65**, 453.

Stacy, R. W., and R. M. Peters. (1965). *In* "Computers in BioMedical Research" (R. W. Stacy and B. D. Waxman, eds.), Vol. II, Chap. 2. Academic Press, New York.

Wald, A., D. Jason, T. W. Murphy, and V. D. B. Mazzia. (1967). A digital computer method to calculate respiratory parameters on a least-mean-square basis. *Intern. Conv. Med. Biol. Eng. Stockholm, Sweden.*

Warner, H. R. (1965). *In* "Computers in Biomedical Research" (R. W. Stacy and B. D. Waxman, eds.), Vol. II, Chap. 10. Academic Press, New York.

Welkowitz, W. (1964). Postoperative monitoring system. *Ann. N.Y. Acad. Sci.* **118**, 400.

CHAPTER 9

Computer-Based Patient Monitoring

HOMER R. WARNER

DEPARTMENT OF BIOPHYSICS AND BIOENGINEERING, UNIVERSITY OF UTAH,
LATTER-DAY SAINTS HOSPITAL, SALT LAKE CITY, UTAH

I. Introduction 239
II. The Monitoring System 241
   A. Monitoring Sequence............................. 243
III. Summary.. 249
    References 251

I. INTRODUCTION

The concept of equipping special areas within a hospital to care for patients whose status is considered either critical or unstable is widely accepted. Such intensive-care units have been established for patients with recent myocardial infarctions, patients recovering from major surgery of various kinds and patients in shock due to trauma, burn or other causes. To be effective, such a unit not only must be a concentration of specially trained medical and paramedical personnel but it must also provide some devices that aid these persons in detecting the true physiologic state of the patient on a continuous basis and that present this information in a form that makes it possible for optimal therapeutic decisions to be made. In this chapter an intensive-care system will be described that incorporates a general-purpose digital computer as an integral part of such a patient care system.

A system has been developed at the Latter-day Saints Hospital in Salt Lake City for computer-based monitoring of patients in a six-bed intensive-care ward to which patients return after open-heart surgery and remain for one to ten days postoperatively. Because many of these patients have undergone lengthy operative procedures with multiple blood transfusions, implantation of artificial heart valves, or other major readjustments of circulatory dynamics produced by surgical correction of intracardiac defects, complex physiological adjustments are encountered for the first few hours or days following such a procedure. During this time, control of the internal environment of the patient is, to a large extent, the responsibility of the medical-care team, since many of the physiological control mechanisms are themselves undergoing major adjustments. To assist in such therapeutic decisions the physician and nurse must be constantly aware of certain key physiological variables, particularly those that are related to the circulatory, respiratory, acid–base, and fluid-control systems.

To accomplish this information-gathering task a system has been designed to provide first a direct analog data gathering mode, which operates continuously to provide oscilloscopic display at each bedside of the electrocardiogram and central arterial pressure waveform as well as a heart-rate meter read-out that has an alarm system to alert the nurse should a drastic change in heart rate occur. This first line of defense operates independent of the computer to detect serious changes in patient status due to such catastrophic events as ventricular fibrillation or cardiac arrest.

The second function of the monitoring system is to derive an estimate of certain basic parameters of physiological performance indirectly from measurement of other variables in the system. Specifically, calculations are made from the shape of the central arterial pressure wave that permit accurate estimation of stroke volume, heart rate, cardiac output, mean pressure, duration of systole, peripheral resistance, and systolic and diastolic pressure on a beat-by-beat basis. The derivation of the equations used for this calculation and the empirical validation of the method against direct measurements in dogs and humans has been reported elsewhere by the author. (See Warner (1968) and Warner et al. (1968)).

Another role the computer plays in the monitoring system is that of screening data derived in this fashion, using statistical methods in order to bring to the attention of the doctor or nurse only that information which represents a statistically significant change in the patient's status since the last data recorded. This function, of course, reduces the amount of information that must be scanned by those making decisions and allows them to concentrate their attention only on those data that contain new information.

In spite of the fact that this approach limits the amount of information presented to the medical personnel to that which is statistically significant, the persons using the data must still make the difficult decisions as to the physiological significance and therapeutic implications of the information presented. To assist in this process, computer programs have been written that allow the nurse or doctor to review the data collected in a variety of ways in order to place the new information in context of other data and to visualize the time sequence leading up to the present state. In addition, programs operate in the system that allow storage and retrieval of clinical information that is essential for subsequent interpretation of these physiological observations.

In the remainder of this chapter each of these functions of the computer-based monitoring system will be described in some detail, following a short description of the computer hardware configuration and time-sharing monitor system under which the computer operates. The system has been in use since June 1966, and a description of some of our experience with it will be presented.

II. THE MONITORING SYSTEM

The MEDLAB time-sharing monitor system was developed at the Latter-day Saints Hospital for the Control Data 3200 computer in order that medical research workers in several laboratories could access the computer on demand from their own laboratory or office to process physiological data, test mathematical models, and debug programs (Pryor and Warner, 1966). Such a system is necessary in this kind of environment if the computer is to be used effectively, since very often the time required to sample the data is long compared to the time required for the computer to carry out the necessary calculations. Yet the computer must be available to do this sampling when required by the experiment.

Under the MEDLAB system, up to four users may be actively sampling data at any one time. Each user specifies his own sampling rate and multiplexing sequence for his analog channels and need not be concerned about the activities of the other users. All data are entered into the system in a nonbuffered mode, with each sample entered into the arithmetic register and handled directly by the appropriate program when sampling occurs. No user may disable the interrupt more than 100 μsec to process a particular sample. Computing is continued in any one program to completion, to a point where additional data or operator intervention is required, or until the program is interrupted by an external or sampling-clock interrupt from another program.

A second identical computer was added to the system one year ago to handle the increasing load. This machine operates on a different version of the MEDLAB monitor which permits 12 user programs to be in core memory at any one time but does not permit compilation or debugging of programs. Programs are called from a remote station into either computer using a four-digit octal switch to specify the program number (Fig. 9-1). Both computers access a common disc controller and set of Control

Fig. 9-1. Remote computer console built around Tektronix memory oscilloscope.

Data 854 disc drives, which are used for storing programs and data. The MEDLAB monitor consists of a set of reentrant subroutines available to the user and occupies approximately 5000 words of core memory. Each real-time user program may occupy up to 2000 words of core memory. It is easy for these user programs to generate multiple overlays in order to accomplish complex operations using relatively small amounts of core memory. Data files on the magnetic discs are arranged by patient and accessed through the patient's hospital number or bed number. These are linked files of variable lengths, and each entry is time and date coded.

A. Monitoring Sequence

On the day prior to open-heart surgery the patient is brought to a small laboratory equipped with a remote computer terminal, an examining table, and a specially designed armboard[1] that permits hyperextension of the patient's wrist and immobilization of the arm for introduction of the monitoring arterial catheter. Two technicians perform the procedure. After introducing a small amount of procaine in the skin over the radial artery, a thin-walled 18-gauge needle connected to a strain-gauge pressure transducer by way of the special catheter[2] filled with saline, is introduced percutaneously into the radial artery. The pressure being transmitted through the needle is visualized on an oscilloscope as the needle is advanced. When the needle enters the artery, the arterial pressure waveform appears on the oscilloscope and the technician advances the catheter through the needle until the tip lies in the subclavian artery. This Teflon catheter is 100 cm long and $\frac{1}{2}$ mm i.d., and is housed in a solid plastic sheath that permits the technician to manipulate the catheter without contaminating it. This eliminates the need for sterile gloves, and draping the patient's arm and allows the whole procedure to be completed in 15 min.

When the catheter has been advanced blindly into the subclavian artery, control measurements are made. A cardiac output value determined at a prior diagnostic heart catheterization or arbitrarily chosen is entered into the computer. Calculations of the required parameters from the central arterial pressure wave are then made and a constant determined that will be used for all subsequent calculations of cardiac output by the pressure-pulse method. In some patients a dye dilution curve is performed and analyzed on-line in the computer prior to introducing the catheter through the needle by injecting dye into an antecubital vein. This is not done routinely, however, because in most instances it is the change in cardiac output that is of importance for subsequent decision making, not the absolute value in a particular patient. After the control measurements have been performed, the catheter is disconnected from the strain gauge, filled with a heparinized saline solution, dead-ended, and strapped to the patient's arm. The needle is withdrawn over the catheter and pressure is applied over the artery for about two minutes to prevent bleeding around the catheter. The patient is then returned to his room to await surgery the next morning. Over 500 such procedures have been performed to date by technicians without complication.

The next morning, the patient is brought to the operating room, where

---

[1] SAFLEX, Romney Engineering & Manufacturing Co., Salt Lake City, Utah.
[2] 310-018 CAP Infusors, Sorenson Research Co., Salt Lake City, Utah.

the catheter is connected to a P23D Statham strain-gauge pressure transducer which is strapped to the patient's forearm. Using a remote computer console at the side of the operating table, the anesthesiologist calls the program and selects the calibration option. On request from the computer, he sends a zero and then a 100-mm Hg pressure calibration. From this point on he may obtain measurements of stroke volume, heart rate, cardiac output, mean pressure, duration of systole, systolic pressure, and diastolic pressure on demand by pressing an interrupt button on the console. Also he may enter codes to log clinical events in the patient's computer-based record along with the physiologic data as the surgery proceeds (Warner et al., 1968). Other options are also available to him, which will be described below in connection with the intensive-care ward.

At the completion of surgery the patient is taken to a six-bed intensive-care ward. At the head of each bed are connections for two pressure transducers and an ECG lead. The three signals sent to the computer include the central arterial pressure wave obtained through the catheter placed preoperatively, a central venous pressure wave transmitted from a second strain gauge connected to a small catheter whose tip lies in the superior vena cava and which was placed by the surgeon at the time of operation, and the electrocardiogram. From the central venous pressure signal the computer determines the mean venous pressure over each heart cycle and the respiratory excursions in thoracic pressure from the fluctuations in this mean pressure, and from the same signal, the respiratory frequency. A single ECG lead is also monitored to give an independent measure of heart rate in the computer. Any discrepancy between the heart rate measured from the central arterial pressure wave and that measured from the ECG may be due to a pulse deficit or to an artifact in the measurement. Such a discrepancy, of course, is brought to the nurse's attention.

When the patient arrives on the ward, a calibration of pressure gauges against a mercury manometer is repeated. Both catheters are filled with heparinized saline and are flushed only if the waveforms become damped. Detection of changes in waveform due to damping is simplified by an option in the program that permits the operator to overlay the last-recorded central arterial pressure waveform with a previous waveform. This makes it easy to detect any loss in high-frequency component at the dicrotic notch. If the catheter manometer system has no leak and no air bubble, high-fidelity pressure recordings can be made for several days without flushing. More often, however, a flush is required approximately every eight hours.

Certain clinical information is entered into the patient's chart when he is admitted to this six-bed ward. First a code is entered to indicate the

9. Computer-Based Patient Monitoring

time of admission. The patient's weight and temperature must be entered, since this information is used in calculating insensible water loss. The present status of intake and output must be updated from surgery so that the continuous record of blood, intravenous solutions, urine output, and water-seal drainage maintained during the surgical procedure by the anesthesiologist is kept current. The patient is assigned a bed number so that all subsequent reference to his record may be made by referring to this number while he is on the ward.

To initiate any action in the computer the nurse, doctor, or technician presses the CALL button on the console (Fig. 9-1). This calls the patient-monitoring program into memory and writes a message on the memory scope requesting the bed number of the patient in question. The operator presses the appropriate numbered key on the front panel of the console and the computer responds by writing the patient's name and age and a list of options as shown in Fig. 9-2. If the operator chooses option 2, a schedule of measurements will be initiated on this patient. The computer will sample the three waveforms 200 times per second until 64 heart beats have been detected.

From each heart cycle the following variables will be determined: stroke volume, heart rate, cardiac output, duration of systole, peripheral resistance, systolic pressure, diastolic pressure, mean central venous pressure, respiratory rate, and respiratory amplitude, along with an independent measure of heart rate from the ECG. From each of the first seven variables a mean value and a standard error of the mean is determined as shown in Fig. 9-3. This is the so-called baseline measurement against which subsequent measurements will be compared to determine whether a statistically significant change in patient status has occurred since the baseline measurement.

This comparison is based on two criteria. First, if any subsequent mean value over 16 heart beats of any variable differs from its expected value by more than three standard errors, a red light is turned on corresponding to that bed to notify the nurse that a change in status has occurred. A second criterion detects a trend. If, on three successive measurements, the mean over 16 heart beats differs by more than one standard error from its expected value in the same direction, a warning light is turned on to alert the nurse that a significant trend has been established. Thus this alarm system takes into account the inherent variability in each variable on a particular patient and does not require the nurse or doctor to enter an arbitrary limit.

The scheduling of measurements is controlled by the program and is based on the past behavior of the patient. After initiation of a schedule, the next reading of 16 heart beats is made four minutes after the baseline

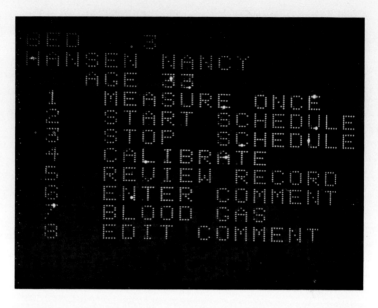

Fig. 9-2. Photograph of memory-oscilloscope screen showing a list of options provided to the doctor or nurse upon calling the program.

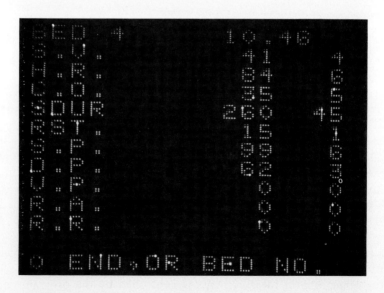

Fig. 9-3. Baseline values showing the mean and standard error of the mean along with the time of the measurement.

9. Computer-Based Patient Monitoring 247

measurement. If this measurement is not significantly different from the baseline, the next scheduled reading will occur eight minutes later on this patient. If the next measurement is within tolerance, the sampling interval is doubled again to a maximum of 16 minutes. If, however, any measurement results in a warning light to the nurse, the sampling interval is immediately reduced to two minutes. Since the sampling interval is known, it is not necessary to save data that does not differ significantly from the baseline value, since such data contain no new information. Only baseline values and measurements that result in turning on the warning light are saved in the patient's computer-based record. This greatly simplifies the job of reviewing the patient's record and allows the reviewer to focus his attention on those data that contain significant information.

Option 1 allows the nurse or doctor to obtain a measurement, on demand, of the physiologic state of the patient. Under this option every other heart beat is plotted back on the oscilloscope as shown in Fig. 9-4 with three vertical marks to indicate the points on the waveform detected by the program as the onset of systole, the peak systolic pressure, and the dicrotic notch. This display serves to confirm to the operator that the pattern-recognition portion of the program is working properly. This is particularly important, since all calculations made from the waveform depend on detection of these key points.

Option 5 allows the doctor or nurse to review the computer record on his patient. Some of the review options are listed in Fig. 9-5. Option 1 permits a page-by-page review of the time sequence of all variables, beginning at any arbitrary point in time and moving either forward or backward. With each display is shown the values for each variable at that point in time alongside the values for the corresponding variable measured at the time of the baseline measurement. Physiological variables may also be reviewed, one variable at a time, with ten values on each page of the review and the times of the measurements listed in a parallel column. Option 3 permits a plot of the time course of one or more variables over a time period specified by the operator and scaled according to options provided to the operator. Option 4 provides a plot of the central arterial pressure waveform recorded at the time of the last measurement and permits the operator to obtain, superimposed on this, a plot of the pressure waveform at the time of the last baseline measurement. This will often give a clue as to the nature of the physiologic disturbance.

Review option 5 allows review of the fluid intake and output status of the patient. The computer requests the number of hours over which the review is to be performed and then asks the operator to enter the amount of fluid remaining in bottles of I.V. solution currently running and the

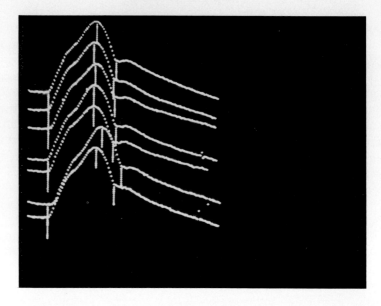

Fig. 9-4. Waveforms plotted during the calculation of stroke volume from the central arterial pressure wave showing the points of pattern recognition as short vertical lines.

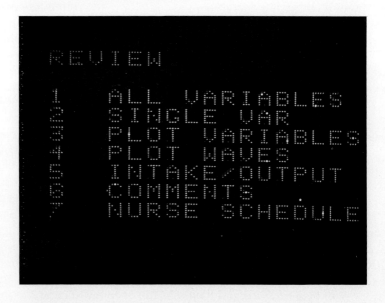

Fig. 9-5. Options available for reviewing patient data.

9. Computer-Based Patient Monitoring 249

amount of blood remaining to be given from blood that has already been logged into the patient's record. These amounts will be subtracted from the amount logged in over the period of time being analyzed. Then the amount of urine collected since the last urine output was logged and the amount of water-seal drainage collected is requested to be entered by the operator. In response to this information, the message shown in Fig. 9-6 is presented showing the total input, the total output, and the net fluid balance on that patient. Sweat or insensible loss is calculated, based on the weight of the patient, his temperature, and the number of hours over which the analysis has been made.

Clinical information, in the form of nurses' notes, can be entered into the patient's medical record using a four-digit code. These four-digit codes are arranged to form a tree structure in which the first digit indicates general categories, as shown in Fig. 9-7. The operator may find the item to be entered by working his way down the tree, one level at a time, or may enter the code in its entirety, or just the beginning digits if he knows the general category into which he wants to proceed. Thus the system of codes is self-teaching and the personnel (both anesthesiologists and recovery-room people) quickly learn the common codes and can enter the information directly in a short time. Upon entering the code, the interpretation of the code is displayed on the scope for confirmation, and additional information, such as drug doses or composition of I.V. fluids, can be added. Information can be logged back in time, if requested; otherwise, the time of the entry is logged automatically into the record.

Review option 6 allows the operator to enter a comment code and review the time sequence of all entries under this code number. Option 8 on the original option list provides a means for editing all comments entered into the record.

The system also provides the nurse with a means for scheduling her own activities. For instance, if a drug is to be given at regularly scheduled intervals, she may, by entering the code number of this drug and choosing a schedule option, specify the interval between measurements and the time of the first measurement. When time for a nurse activity occurs, the yellow light is turned on; the nurse presses this yellow light and is informed as to what activity is due. If she is ready to perform that activity, she merely presses the appropriate option button, which causes the activity to be logged and the schedule to be updated.

III. SUMMARY

Thus the system in its present state of development is capable of assisting the patient-care personnel in several vital ways in caring for

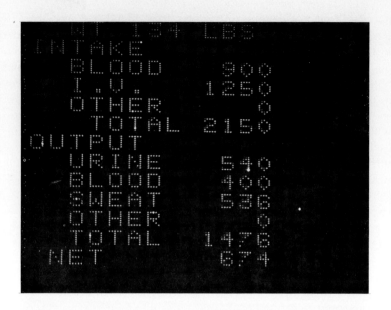

FIG. 9-6. Report of the fluid intake and output of a patient.

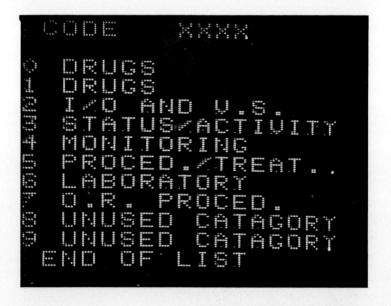

FIG. 9-7. First page of nurse comment directory.

these critically ill people. First, considerable new information is provided in the form of in-depth measurement of variables difficult to estimate by indirect methods and essential for some of the decision making that must take place in this kind of environment. Second, the physiological measurements made are screened so that only those that, by statistical criteria, represent a change in a patient status are presented to the nurse or doctor for decision making. Third, a complete log of clinical, as well as physiological, activities is maintained in the computer and can be accessed in a variety of formats that allow interpretation of interrelationships not obvious otherwise and that will be helpful in decision making. Although other options in the system, such as blood gas measurement, are being introduced as the need arises, the basic elements of the system already play a vital role in patient care under these circumstances. If, for some reason, the system for monitoring is inoperative, most surgeons will now cancel surgery until the monitoring is available. This is perhaps the most telling evidence that such a system does, in fact, play an important role in improving patient care.

References

Pryor, T. A., and H. R. Warner. (1966). Time-sharing in biomedical research. *Datamation*, **12** (4), 54.

Warner, H. R. (1968). Experiences with computer-based patient monitoring. *Anesthesia Analgesia* **47** (5) L, 453.

Warner, H. R., R. M. Gardner, T. A. Pryor, and W. M. Stauffer. (1968). Computer automated heart catheterization laboratory. UCLA Forum in Medical Sciences.

Warner, H. R., R. M. Gardner, and A. F. Toronto. (1968). Computer-based monitoring of cardiovascular functions in postoperative patients. *Circ. Suppl.* **38,** 68.

CHAPTER 10

The Comprehensive Patient-Monitoring Concept

RALPH W. STACY[1]

DEPARTMENT OF SURGERY, UNIVERSITY OF NORTH CAROLINA, CHAPEL HILL,
NORTH CAROLINA

I.	Introduction	253
II.	The Nature of Death	254
III.	Crisis-Prevention Interference	262
IV.	The Real Purpose of Patient Monitoring	263
V.	The Requirements of a Patient-Monitoring System	265
VI.	The State of the Art of Patient Monitoring	268
VII.	A Practical Comprehensive Patient-Monitoring System	271
VIII.	Summary	274
	References	275

I. INTRODUCTION

During the past few years, a number of developments in biomedical computing have created hardware and software systems that now make it possible to carry out the monitoring of acutely ill patients on a much more extensive and useful basis than has ever been done before. The chapters immediately preceding this one bear witness to the rather complex technology that has grown up around this concept. Although computerization of patient monitoring is still in an infant phase, the effort promises to bear fruit of a sort that will be unique in medical science of this age.

This chapter is written after the editors' having read and evaluated the other chapters of this volume on the subject of patient monitoring. Its intent is to fill a gap that has occurred as a result of the other writings being carried out by separately thinking and working scientists. We justify this after-the-fact writing on the basis that a good editor, having detected a fault in the volume on which he is working, will make an effort to correct

[1] Present address: Cox Heart Institute, Kettering, Ohio.

that fault. It is our intention that this chapter shall complete the current thinking on patient monitoring with computers, and shall present those basic concepts which underly the entire effort, but which have not been expressed in the preceding pages.

Any research effort must have as its ultimate goal the application of its findings to the welfare of the human race; sometimes this step is a large and difficult one, and sometimes the application of information must await technological developments that do not exist at the time the knowledge becomes available. In this case, however, we believe that the ability to carry out the ultimate in patient monitoring—the operation of an intensive-care unit that utilizes computer techniques to weld the many facets of the ICU into a system that will prevent the death of individuals—is in fact in existence and can be assembled. There is much work to be done, but the intellectual capabilities are at work on the problem and the task will be done.

We will here present a thesis made of several hypotheses, each of which can be defended and developed. In logical order, these hypotheses are:

(1) That death of the human individual is tantamount to catastrophic failure of an artificial system, and the processes of death are by nature parallel to the processes of such system failure.

(2) That it is possible to derive predictors of the onset of crisis, and with the knowledge of such predictors, it is possible for man to interfere with the process and thus prevent death.

(3) That the real purpose of patient monitoring (particularly in the ICU) should be the gleaning of information from which such predictors can be derived.

(4) That we have sufficient knowledge to define the requirements of a computer-based monitoring system that will be capable of carrying out the above.

(5) That with modern, third-generation computers and the accessories that have been developed for them, we have the hardware capability to do this within an economic structure that can make the system available generally to man.

(6) That the ultimate success of a patient-monitoring system will depend to a large extent on the comprehensiveness of the concepts on which it is based—a monitoring system that provides too little information can be of only limited usefulness in the task at hand.

II. THE NATURE OF DEATH

It is rare indeed that one thinks of death of the individual as being anything other than biological—a cessation of life processes that is accepted as irreversible and representing the end of life. There is, however, another way of thinking of death, one that may in the final analysis lead to much greater understanding of the factors that combine to bring about the cessation of life. For such thinking, we must turn to those methods which have been developed in large-scale technology for the evaluation of missile guidance, communications, and other complex man-made systems, and for the prediction of failure of such systems.

In engineering, the failure of a system's function that results in the total destruction of the system is referred to as "catastrophic failure." We can see no difference between the failure of a space probe which causes that probe to flounder wildly into outer space, beyond recovery, and the death of a human individual.

We have long been accustomed to thinking of the "nature of life," and we do indeed know a great deal about the basic processes that go into the pattern of behavior of the organism we call "biological living." Is it then difficult to proceed from this to logical analysis of what must go on when the system becomes incompatible with the continuation of function?

The fully developed human organism is a complicated system, consisting of a number of subsystems so integrated that they function as a whole to assure the continuation of exchange of vital materials with the environment to maintain the thermodynamic equilibrium called "life." Each of these subsystems is in itself a complicated system, and the analysis of these systems as cybernetic units has only just begun. Even so, the *basic* nature of the physiological systems is well known to us. It has been, in fact, since the writings of Claude Bernard more than a century ago.

Every one of these subsystems is a negative-feedback control system. Every one of these systems, when subjected to a stress, undergoes some deviation from its control level, and if it is a viable system it automatically generates changes that restore the original level. The cardiovascular system, the respiratory system, the acid–base regulating system, the fluid balance system, the temperature regulating system—all have the same basic nature. A number of these systems have been subjected to close scrutiny, and the literature now contains numerous writings in which the mathematics of such systems has been carefully dissected (Grodins, 1963; Coulter and Updike, 1965).

But there has been little or no attention paid to the functioning of these systems in symphony, a requirement for the continuation of life of the *whole* organism. It is time we looked at this synthesis of subsystems into the larger organism and examined its capabilities for continuation of viability.

At this point in time our approach to the system existing at the organism level must of necessity be somewhat artificial; it will be revealing even so.

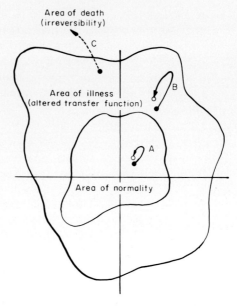

Fig. 10-1. A phase plot of two variables that may be used to locate the individual's condition. (a) represents the response to stress when the individual is in the normality space, (b) when the individual is in the space of abnormality but viability, and (c) when the application of stress carries the individual outside the area of viability. Real systems would exist in multidimensional space, of course.

As a beginning let us examine the diagram shown in Fig. 10-1. This is simply a two-dimensional plot, whose coordinates might represent only two of many descriptors of the process of life; hence, it is a gross oversimplification of any scheme that might be real. Still, using such a diagram, one can point out certain salient features that will shed light on the thesis we are developing.

First, let us assume that the *completely* normal individual would exhibit characteristics with respect to these two coordinates that would place him in the diagram at the center of the plot. His condition could then be described as a point of a diagram, which point would be the absolute ideal. Now we all know that no two humans exhibit exactly the same physiological condition; therefore, it is obvious that we cannot describe normality as a point in a space, even in two dimensions. It is in turn apparent that if we are to arrange this description to include more than one individual,[2] we must designate "normality" not as a point, but rather as an area. And furthermore, because we cannot define normality in precise terms, we

---

[2] Or, for that matter, any one individual at different times.

must realize that the perimeter of this area is not a clean line of demarkation but a gray area, in which some small part of the population may fall and still be called "normal." Indeed, our knowledge of the statistical distribution of physiological parameters would lead us to the conclusion that a plot of individual positions would result in a cluster about the center of the diagram, with decreasing point density occurring as one moves away from the center.

The definition (physiologically) of the center of this diagram is in itself a major physiological problem, for we do not know the factors that determine the center (mean) level of control of the cybernetic biological systems. A most interesting set of concepts concerning this determination of control-system levels has been conceived and described by Coulter (1968); unfortunately, it would seem that completion of this analysis may be delayed by lack of sufficient information or even by lack of the mathematics required to handle the problem. For our analysis we may proceed simply by making the assumption that we can in fact describe our coordinates in terms of the statistical mean of observed physiological parameters, so long as we continue to be aware of the probabilistic variations inherent in the scheme.

Within this area of normality, any of the physiological subsystems under discussion will, when subjected to stress, undergo a shift in its position in the descriptive grid, and this shift is such that the position will deviate *away* from the original position. The amount of shift will depend on the severity of applied stress, the time lag associated with the transfer function of the control system, etc. However, because the system is normal, it will by nature return toward (but not entirely to) the original position. The permanent shift that remains after the return is complete is a measure of the error function of the control system. If and when the stress is completely removed, the system would presumably return to the original position *unless permanent damage has occurred* which has resulted in the shift of "normality." The response of the normal system is diagrammed in Fig. 10-1 as a solid loop (for pedagogic purposes).

The degree of shift, both transient and permanent, associated with the application of a stress can be described by the transfer function of the normal system only so long as the position held by the system remains in the area of normality; the stress may then push the system into a new area surrounding the area of normality, which we must now call the "area of illness." The system is no longer normal—its transfer function is altered in some way about which we know very little. The interaction of this now abnormal system with other systems must also be abnormal, with the result that the entire organism is "sick." Unless restorative measures, either natural or man applied, are brought into action, the individual will remain sick and will not return to the area of normality.

In most cases, of course, body defenses (with the aid of therapy, surgery, etc. in many cases) do in fact remove the stress factor and permit the admittedly altered control systems to function to the degree that the system is restored to the area of normality and the recovery from illness occurs. It is not these cases with which we are concerned; it is those in which the deviation away from the area of normality is of such degree that there is a real danger of failure of the systems, even with the help of man, to ever return to normality.

Still working within our hypothetical two-dimensional space, it is apparent that we must now define yet another perimeter, representing the outer limits of the area of illness, which with the enclosed area of normality might now be termed the "area of viability." Should the starting position of the system descriptors be near these limits of viability, and should the applied stress be great enough or the transfer function altered enough that the position is shifted by the stress outside the area of viability, the control systems lose all their restorative power; they may even become positive-feedback systems, in which case they will go into wild gyration, out of control, and the catastrophic failure of which we have spoken will occur. The human dies.

Again, it is apparent that the outer limits of viability in any physiological system cannot be described in clearcut numbers. This is also a gray area, with statistically diminishing populations as the position within our grid varies further from its gravitational center.

Thus, in the coldly logical sense, death is a matter of alterations from normality of the starting position of a system subjected to stress, and the ability of the sytem to respond to the stress is in itself a function of that starting position. Death occurs when a complex situation is generated in which the combination of starting point and response to stress carries the system outside the area of viability.

Mathematically, it is impossible to make more than a crude beginning in the description of this situation. The model we must use is perfectly clear, however, and we can in fact derive considerable understanding of the nature of death and the functions that must be carried out to prevent it by examining even the meager mathematics we can generate now.

First, we must recognize that there are many systems that go into the total "organism system," and that each of these is interrelated with all of the others in complex ways only dimly understood at this time. For any one of those systems, one could describe a "condition" or "status position" or "operating state" in general terms as

$$Y_t = G(H, S, t)$$

where Y_t is the status position at time t after application of stress, G is the

10. Comprehensive Patient-Monitoring

transfer function of the system as it exists at the time of application of the stress, H is the initial status position (prestress), S is the stress, which varies in magnitude and with time, and taking into account that there may be n systems involved in the organism, and using the subscript (0) to refer to the total organism,

$$Y_{0_t} = G_1(H_1, S_1, t) + G_2(H_2, S_2, t) + \cdots + G_n(H_n, S_n, t)$$

It should be pointed out here that we do not now know even the nature of the operators in such an equation; these may in themselves be complex relationships. Actually, there is every likelihood that the transfer function G of any system may change during the response to stress, so that at time t following the application of stress, the total system status position Y_{0_t} would probably be like the following:

$$Y_{0_t} = K_1 G_1(H_1, S_1, t) + K_2 G_2(H_2, S_2, t) + \cdots + K_n G_n(H_n, S_n, t)$$

where K is a second function describing the variation of the transfer function G with time progress of the response to stress. We do not present such equations as actually descriptive of the status position of the organism; we are merely saying that the mathematics of the situation is complex and will take some form like that shown.

These descriptions tell us something about the initial position of the system in its function space and about the response of that system to stress.

The equations presented above are, of course, very much lacking in information content, because they do not tell us of the real relationships between the many variables included, nor do they tell us the identity of the variables. In order to carry the mathematical description of the entire complex to a degree of usefulness, it is most likely that we will ultimately resort to the use of newer methods of mathematics, including especially the synthesis of *interaction matrices* of the various interrelated systems. This very interesting tool of mathematics is called *multivariable analysis* and is to be distinguished from but is closely related to the statistical procedures of multivariate analysis. In the case of multivariable analysis, relationships between various subsystems in a whole system are described by quantities that appear in terms falling to either side of the diagonal of the interaction matrix.

Control-system theory is in fact mushrooming in this direction at the time of this writing, and it is highly likely that the mathematics required for application of these techniques to the analysis of death processes already exists. If not yet, then the intensive research now going on in these areas should result in the mathematics being ready for use within one or two decades, which time will be required in any case for acquisition of the basic data required for its use in analysis of death.

One of the tendencies in the analysis of such systems is to define the system in terms of "state variables" and proceed with the analysis in terms of what happens to these state variables as each is changed in turn.[3] In essence, this is quite equivalent to the analog-computer solution of a complex system of differential equations.

Thus the state of the art in system analysis is just entering a phase that will be most useful to the medical scientist interested in the processes leading to death. The greatest problem with which this scientist will be faced is one of judgment—the selection of the mathematical tools to use to accomplish his purpose.

It is very apparent from the sketchiness and looseness of the preceding description that the thesis we are developing here is one of concept only; the actual data on which the projected analyses are to be based and the tools to be used in such analyses remain to be obtained and collated into useful channels.

For the next step in logic we must now turn to an estimate of the *probability* that response to stress will carry the system beyond the limits of the viability space. Actually, this will probably be most useful as two separate probability estimates: (1) an estimate of the probability that the organism will enter into a "crisis" situation in which there exists a possibility of catastrophic failure, and (2) an estimate of the probability that the currently predictable stress will result in catastrophic failure and produce death.

The descriptor of the probability of entry of the organism into crisis P_c may take the form

$$P_c = L(Y_{0_t}, S)$$

where L is in itself a probability function describing the likelihood of the system's responding to stress in such a fashion as to change Y_{0_t} toward the limits of viability enough to produce a finite probability of failure. In such a case P_c is then a probability of entering into the "crisis situation."

Once the organism is in crisis (i.e., when P_c assumes significant magnitude), there is then the possibility of deriving yet another descriptor P_d, which describes the probability of the organism's reaching the limits of viability when subjected to a stress and the probability of such disastrous stress occurring in the situation in which the organism is existing. This probability of death must be a combination of the probability of crisis, the existing situation point coordinates, the probability of occurrence of stress, and the transfer function describing the situation of the moment. That is,

$$P_d = M(P_c, Y_{0_t}, S)$$

[3] In such analysis both "operating state" and "structural state" should be considered. Our discussion to date has been that of the operating state.

10. Comprehensive Patient-Monitoring

In the final analysis it is likely that there will be a functional relationship between the probability of crisis P_c and the probability of death P_d such that the computation of "situation condition" will involve one lumped probability parameter.

In the intensive-care monitoring situation, it is important that we be able to predict the onset of crisis. In some other cases (e.g., in the selection of transplant donors) it is important to be able to provide an accurate and reliable estimate of the probability of death.

Up to this point, in order to handle the concepts at all, we have limited our description to a situation requiring only two dimensions for location of a status position. Actually, the status position of a real physiological system will most certainly have to be described by coordinates placing it in an n-dimensional space. The prospect of trying to describe such a system mathematically, complicated as it is by the dimensional multiplicity and the necessary introduction of probability functions, would appear to render the task hopeless. It is not, however, hopeless—it is merely very difficult.

There may in fact be one redeeming situation in the interaction of systems in the organism that may simplify the problem appreciably. This is that although there are numerous sytems that must remain viable for the continuation of life, biological death is usually associated with the failure of one of those systems. Thus one can say that a death has occurred because of heart failure or respiratory failure. If the heart fails, it follows that the respiratory apparatus will fail soon after, that metabolic processes will fail shortly after the loss of oxygen exchange and blood flow, and so on.

It may well be, then, that in the final analysis, the predictors of crisis or of death may be computed on the basis of the status of only one system; obviously, this would simplify the mathematics a great deal.[4] To be sure, the predictors of a single critical system would also involve the functional status of other systems, so in no event is it likely that the computation of predictors will be very simple.

It would seem at this point that there is no real benefit to be derived from even trying to write these probability functions in such terms as those of the equations above, for we know far too little about the probabilities involved. Medical research to date has not provided quantitative information on this aspect of terminal illness.

Thus we have arrived at a concept of the nature of death. Let us now be absolutely certain that we have not created a misimpression; death of

---

[4] That is, there may be a possibility that "snowballing" might be predicted by estimating the likelihood of occurrence of the initial event causing the snowball to roll.

the organism is nearly certain to be the result of the total failure of function of one critical system. This failure in turn results in the failure of another, that of another, and so on. Thus one could say that prediction of the failure of the most susceptible of the systems would in fact be tantamount to a prediction of death. It must be remembered that all subsystems in the organism are interrelated, and that the condition of the other systems may in fact be contributing largely to the possible failure of the most critical. Death, the termination of life, can be nearly as complex as its continuation.

III. CRISIS-PREVENTION INTERFERENCE

From the above one does not have to stretch his imagination very far to conclude that if we know enough about the responses of systems to stress under all conditions within the viability space, it would be possible to make predictions as to whether any one individual were likely to enter a crisis situation. We are impressed with the observation that with the continuing development of defibrillators, pacemakers, respirators, artificial kidneys, acute-care procedures, and emergency techniques, in the state in which the art of medicine exists today, the *onset of crisis could be prevented if one knew that that crisis was about to occur* (this applies to many situations but obviously not to all).

Furthermore, with the very high probability of solution of the tissue-rejection problem in the near future, transplant techniques will permit the replacement of organs in systems whose probability of failure is high and is based on organ degeneration. All of these developments combined give promise of the greatest advances in medical science of all time, and these advances are likely to occur within the next decade or two.

Crises develop in the intensive-care situation under conditions that are too complex for the medical staff to control; usually this means that deterioration of condition has proceeded too far, or that the physiological situation is so complex that sufficient data are not available and the failure to recognize impending crisis is due to ignorance (admittedly unavoidable).[5] Because the physiological systems are so complex and their interrelations so numerous, it is manifestly apparent that the human needs the mechanical aid of the digital computer.

Those individuals now working in the area of computerized intensive-care research bear the responsibility for making certain that their research will produce the results called for in this very preliminary analysis of the processes of death.

[5] The examples given by the authors of Chapter 7 illustrate this very well.

10. Comprehensive Patient-Monitoring

IV. THE REAL PURPOSE OF PATIENT MONITORING

The concepts outlined above imply that if one can obtain enough information about the condition of a patient and the ability of the patient's physiological systems to respond to stress, he can calculate a probability of entry of the patient into crisis or, in the crisis situation, can calculate the probability that the patient will in fact die. If there exists *enough information* to make such predictions, then there must exist enough information to identify the conditions that are producing such a situation. It should be possible by correction of these conditions to prevent the onset of crisis and of death. This is not only the ultimate goal of patient monitoring, it is the ultimate goal of the entire process of patient care.

An examination of the equations and concepts described above (crude though they may be) reveals immediately that the amount of information now available to the physician for the care of his patient is only a fraction of that actually needed to perform the patient care *intelligently*. If there were not developments occurring that are likely to correct this lack of information, then the dream of being able to predict and prevent the onset of crisis and death would be hopeless. Fortunately, there are such developments, and most important among these is the ongoing research on the computer monitoring of patients in the intensive-care ward. Only by such monitoring, carried to its ultimate degree of usefulness, can our final goals be achieved.

But the achievement of a state of the art allowing for the gathering, digestion, and utilization of information of the magnitude required by the analyses indicated above means that there must be a reorientation of the procedures used in patient monitoring, and of the goals of the monitoring process. Patient monitoring must be carried out not only for the immediate care of the patient but also (as a research procedure) for the ultimate generation of a data base large enough to allow the derivation of the needed predictors. The purposes of monitoring thus become twofold:

(1) Patient monitoring should be carried out to provide enough information on the condition of the patient that it will be possible to give him the best possible care under the existing state of the art.

(2) Patient monitoring should be carried out in such a way that data obtained on each patient contribute to the data store, which will eventually be used to create a new, logically based monitoring procedure. Scientists involved in such monitoring should be continually working out transfer functions under all conditions.

The second of these purposes of monitoring is a research objective, and experience has proved many times that such objectives are very difficult to achieve in a clinical situation. Therefore, the planning and design of monitoring schemes generated and operated in the next decade or two should be deliberately aimed toward creating such a research facility at the same time the patient-care facility is brought into being. It is here suggested that funding agencies and the planners of medical research activities should have this special purpose of patient monitoring in mind at all times.

This consideration (i.e., the difficulty of achieving research objectives in the patient-care situation) suggests also that there should be carefully planned and executed research projects aimed specifically at gathering data on the behavior of system transfer functions under all conditions, using laboratory animals in situations that duplicate as nearly as possible the processes leading to death of the human. Although there are species differences that can create errors in transfer of such information to the human death study, those differences can be kept in mind and appropriately accounted for in the final assembly of the predictor-description mathematics.

All of the above leads us to one other inescapable conclusion: that *the monitoring process will be of maximal use if and only if that process includes enough measurable or computable quantities to allow for assessment of the physiological condition of all the critical systems of the body*. We here define what we shall call "comprehensive patient monitoring" as monitoring in which enough information is gathered to allow for assessment of the circulatory, respiratory, acid–base control, metabolic control, kidney and water-balance control, and central nervous systems. These are the physiological systems the authors now consider to be critical in the survival of the individual in the crisis situation. Other physiological systems (e.g., the musculoskeletal, the endocrine, the sensory, etc.) are essential to the carrying on of life as a whole, but not to *immediate* survival of the acutely ill patient. Our views on this may change, of course, as we accumulate and assemble more information on the system interrelationships.

As a last word in this section having to do with the basic concepts of patient monitoring, let us add the observation that the study of death processes is actually a special case of the study of life processes. These studies, involving as they do the understanding and description of all major physiological systems, mean that when one is studying life and death processes from this point of view, he is studying at one time all that is fundamental in Systems Physiology. It could be that within a few years, the study of Physiology will be approached through the portal of study of life and death processes. It would be logical if the education of medical students should begin with such study, for the physician of the future will spend his entire career in the continuing study and application of the principles of life, death, viability, and the prevention of catastrophic failure.

V. THE REQUIREMENTS OF A PATIENT-MONITORING SYSTEM

We have already pointed out that a comprehensive patient-monitoring system must be encompassing enough that it will allow assessment of the physiological condition of all the critical physiological systems. At first glance, this statement would appear to be asking for the impossible, for the systems are all complicated in themselves and their interrelationships are such as to create almost inconceivable complexity. However, a reasonable and practical analysis of the situation reveals that in truth the requirements are far less than they would seem to be. In the first place we have learned enough about each of the physiological systems that we can eliminate many possible measurements as having been of use in the research situation in learning about the behavior of the system, but as contributing little to an assessment of the ability of that system to function in the milieu in which it finds itself. Detailed analysis reveals, in fact, that the number of measurable quantities is surprisingly small, that many of the data items that are normally measured as separate quantities can in fact be computed from a few basic measurements, and that in the patient-monitoring situation many measurements can be simplified to make the total monitoring task easier.

The patient-care situation, as it now exists in present-day medical care, often calls for the gathering and accumulation of many data that are extraneous to the purposes defined in this writing. Examination of the patient records coming out of an intensive-care unit will reveal that in fact a large percentage of the voluminous notes and accumulated paper records from such units actually amount to verbal descriptions of the opinions of medical personnel and their justifications for such opinions; the careful noting of exact times of and reactions to routine nursing procedures; written physician orders and evidence of the compliance to those orders by nurses and other personnel, and so on. Such detailed analysis also shows that in many situations information is obtained with measurements that are repeated at intervals far shorter than the time it takes a system to change. Thus there is often a wasteful gathering of too much information, which can fill data cells rapidly and which contributes little to the total information that comes out of the system.

Thus an analysis would indicate that it is perfectly possible to reduce the information flow rate by a significant amount simply by making careful judgment as to the required rate of repetition of measurement.

Also, where computer monitoring is carried out, it is often observed that computer methods are substituted for manual methods, at a tremendous cost of computer time and sometimes at the expense of accuracy and validity. Some such places will be pointed out in the discussion to follow. In such cases it is logical to return to the use of the human in the overall

TABLE Ia

Measurable Quantities Required for Comprehensive Patient Monitoring

Quantity	Input form	Interval between measurements	Required sampling frequency
Circulatory			
ECG (wave analysis)	Analog	15–30 min	200/sec for 5 sec
Arterial pressure	Analog	5 min	100/sec for 5 sec
Central venous pressure	Analog	15–30 min	100/sec for 5 sec
PVC count	Digital	30 sec	
Rate alarm	Interrupt	Continuous	
Cardiac output[a]			
Respiratory			
Air flow	Analog	1 hr	40/sec
Transpulmonary pressure[a]	Analog	1 hr	40/sec
CO_2 concentration	Analog	1 hr	40/sec
O_2 concentration	Analog	1 hr	40/sec
Blood chemistry			
pH	Manual[b]	3 hr	
pCO_2	Manual[b]	3 hr	
pO_2	Manual[b]	3 hr	
Sodium	Manual[b]	12 hr	
Potassium	Manual[b]	12 hr	
Chloride	Manual[b]	12 hr	
B.U.N.	Manual[b]	24 hr	
Hematocrit	Manual[b]	6 hr	
Other	Manual[b]	Variable	
Fluid Balance			
Urine output	Manual[b]	1 hr	
Blood loss	Manual[b]	30 min	
Tube drainages	Manual[b]	1 hr	
Oral intake	Manual	8 hr	
Intravenous intake	Manual[b]	1 hr	
Medications and treatments			
As ordered	Manual	When given	
Body temperature	Manual[b]	1 hr	
System responses	(Cardiovascular and respiratory responses to brief exposure to 2% CO_2 in inspired air)		

[a] May be computed from other primary measurables.
[b] May be automated in subsequent versions.

10. Comprehensive Patient-Monitoring

TABLE Ib
DERIVABLE QUANTITIES IN COMPREHENSIVE PATIENT MONITORING[a]

Cardiovascular
   Stroke volume
   Heart rate
   Cardiac output
   Systolic pressure
   Diastolic pressure
   Mean venous pressure
   Respiratory variations in venous pressure
   Cardiac work
   Systolic duration and contraction velocity
   Central arterial reservoir compliance
   Peripheral resistance
   ECG derivables
   Pressure responses to imposed stress
Respiratory
   Tidal volume
   Expiratory reserve
   Vital capacity
   Lung compliance
   Airway resistance
   Work of breathing
   Alveolar–arterial gradients
   Ventilation–perfusion ratio
Acid-base balance
   Serum bicarbonate
   Base excess or deficit
   Total CO_2
   Acid–base status
   Computed requirements for acid-base adjustment
Water balance
   Fluid-balance status
   Renal-function evaluation
   Blood and seepage loss
   Fluid replacement needs
Metabolic energy
   O_2 consumption and CO_2 production
   Metabolic rate and R.Q.
   Metabolic rate corrected to $37°C$
   Body-temperature trends
Correlative quantities
   Circulatory response patterns
   Respiratory response patterns
   System response correlations
   Prognosis
   Suggested therapeutic measures

[a] In our system, we are using 18 primary measurables. The number of possible derivables is virtually unlimited. We expect to obtain 78 such derivables, giving a total of 96 data items.

monitoring scheme. Indeed, there are numerous reasons that it is important that the human and his extraordinary ability to make comparative judgments (his ability for pattern recognition) be incorporated into the monitoring system.

Tables Ia and Ib provide a summary of those factors that, in the opinion of the authors, are required for the assessment of the critical physiological systems in the intensive-care situation. Included in Table Ia are also our estimates of the relative frequency with which such measurements need be made and the data-sampling rates or modes in which data must be gathered. Comparison of these figures with those to be found in other descriptions of patient-monitoring schemes (see list of references at the end of this chapter) will show that this represents a far more easily realizable data base than some of those proposed by others.

VI. THE STATE OF THE ART OF PATIENT MONITORING

For descriptive purposes the process of patient monitoring may be divided into three phases: (1) the acquisition of primary data, (2) the derivation of secondary data by use of the primary data to calculate parameters that cannot or should not be measured directly, and (3) the assembly of the primary and secondary data into descriptive form and the use of this description for the care of the patient. For clarification it may be pointed out that the acquisition of primary data includes the use of electronic devices with which physiological measurements are made, the manual entry of data by medical and paramedical personnel, and the use of analog electronic equipment to "preprocess" primary data and/or generate alarms for the guidance of the medical personnel in charge of the unit. The derivation of secondary data is entirely a function of computation of new parameters from the primary data, using mathematical relationships that have been derived in physiological research over the past hundred or more years. The final assembly and utilization of those data involves, then, the application (by computer or manually–mentally) of clinical judgment and knowledge to formulate a description of the physiological condition of the system; this step also includes the communication between the computer and physician, an area that has received entirely too little attention to date.

Medical electronics has undergone enormous changes in the past very few years, and these changes are by no means complete. With the introduction of solid-state circuitry and, more recently, with the availability of integrated-circuit modules, which are fast becoming stable, versatile, and inexpensive, it is now possible to assemble electronic circuits that are a fraction of the size and cost of units required for the same tasks a few years ago—units that require far less power, occupy much less space, and are virtually

10. Comprehensive Patient-Monitoring

devoid of adjustments to be made by the medical personnel who use them. Very good examples are provided in the electronics developed at the Latter-day Saint's Hospital in Salt Lake City (see Chapter 9). The electrocardiograph and strain-gage pressure-monitoring circuits developed in that project are small, inexpensive, reliable, and require almost no human attention.

In our laboratories we have continued the development of such electronic devices, seeking for further stabilization, design allowing for the use of components selected at random from the laboratory shelves, and further elimination of adjustments required in the intensive-care unit. It is one of our goals to develop (and we now have prototypes that we are evaluating) circuits that are "universal" in the sense that a single circuit can be used for the measurement and amplification of electrocardiograms, arterial pressures, venous pressures, body temperatures, bed weights, fluid-bottle weights or levels, and so on. Thus, the "inbed electronics" would contain power supplies and only one type of circuit card. The function that card would subserve would depend on which of the connector pins are used in the socket into which the circuit card is inserted. Furthermore, we have found it expedient to design the cards in such a way that they will work the same whether they are inserted "right side up" or "upside down," and all balance and sensitivity controls are mounted on the card itself and are inaccessible to intensive-care-unit personnel.

A routine check of balance and sensitivity by trained electronics people once every twelve or twenty-four hours would seem to be adequate for regular maintenance. Extra cards are stocked in the intensive-care unit, so that if failure of one card occurs between maintenance checks, the old card is pulled out and a new one inserted. Since there is only one card type, the training required for this level of maintenance is minimal, and the possibility of error is virtually eliminated.

Thus, the "inbed electronics" reduces to maximum simplicity. In the intensive-care situation, of course, there must also be means for continuous visual observation of certain variables (e.g., ECG and arterial pressure), and there must be provision for the generation of alarms under some situations. This is carried out in a second electronic unit we call the "bedside electronics." This unit necessarily is more complex: it contains an observation oscilloscope; digital displays of heart rate, systolic pressure, and diastolic pressure; circuits for reading of pressure maxima and minima; a circuit for conversion of QRS complexes to rate data; circuits for analog-to-digital conversion for the digital displays; circuits for alarm generation under conditions selected by physician; and finally, a circuit for detecting and accumulating a count of premature ventricular systoles. With the exception of the last of these circuits, all are well within the state of the art of medical electronics. The only reason we have developed these independently in our own laboratories is so that they will have the format we felt was optimal for the intensive-care situation; none of the commercial units available exactly fitted our requirements for this.

The PVC detector circuit (Feezor and Stacy, 1969) is in the evaluation stage at the time of this writing. The ECG (as amplified by the bedside electronics) is fed into this circuit, which in effect is a delay line allowing for the comparison of the QRS complex with a built-in specially selected impulse response that closely approximates the shape of a PVC, but not that of a normally generated QRS complex. A correlation function is generated, which in the case of the PVC is about three times the magnitude of that generated by a normal QRS complex. This correlation function appears as an analog voltage, and a Schmitt trigger is used to generate a shaped pulse for counting or alarm purposes.

It is significant that this scheme seems to work well with a single-chest-lead ECG, which is in place in any case for visual observation purposes. In some monitoring schemes the continuous analog-to-digital conversion of ECG signals and the continuous searching for PVC occurrences (a major pattern-recognition problem for the digital computer) requires very large amounts of computer time and significantly increases the computer hardware requirements. Our performance of this task at the analog level has reduced the total information-flow requirement a great deal. If in the course of evaluation we find that our existing circuit will not perform as we believe it will, we will continue to search for such an analog method for detection of PVC's.

Thus the analog electronics required for comprehensive monitoring already exists, and will be completely debugged and functional in a very short time.

From Table Ia it will be noted that many of the primary data required for comprehensive monitoring can actually be entered manually, since the data rate is quite low and the manual entry poses no particular problem for medical personnel working in the intensive-care unit. In fact, it is our philosophy that automation can and should be used only when it is possible to carry out a measurement automatically with little chance of failure and with accuracy and reliability at least as good as that which can be obtained by human observation. However, we feel that there are many situations (particularly those involving pattern recognition) where the state of the art is not such that automatic devices can do a better job than the human, and therefore such measurements should not be automated at this time.

Basic research in the programming of computers for calculating secondary data items from primary data input has been ongoing for some years, and it is even now possible to assemble working programs that can be used for derivation of all of the secondary items listed for comprehensive monitoring. Over the next decade it is anticipated that even more such programs will become available, so that computer analysis of physiological condition should become easier and more reliable as time goes on.

The area in which there is greatest need for work is in the computer interpretation of quantitative data and the genesis of system condition reports that are meaningful to the physician and helpful to him in the treatment of the patient. Such reports must be made available to the physician in such form that they can be obtained by him on demand without delay, and in such format that they present the important information without a great many extraneous data that have no significance to the situation in hand. The hardware for generation of reports in a few milliseconds and display of those reports on large-screen television monitors is available and has been interfaced with computers for this purpose. Hard-copy output, when desired, can be obtained by computer printout or by photographically recording the display screen; such hard-copy output is kept to a minimum in our system for reasons of computer printout-time economy and to minimize the problems of permanent record storage.

In summary, the state of the art in patient monitoring is such that it is completely possible to carry out comprehensive monitoring in the intensive-care situation *at this time*. The greatest problems remaining to be solved are those having to do with the mode of transmission of information from the computer to the medical staff, and these must be solved by teams including physicians and computer technologists or by individuals who are both physicians and computer technologists.

The authors recognize that the acquisition of information required for comprehensive monitoring and for the achievement of its ultimate goals will be difficult in the patient-care situation. However, we believe that with diligent effort on the part of the physicians involved and with supplementing of such information as does become available from human studies with carefully controlled laboratory-produced data from animals other than man, we can—we must—in a few years acquire the information needed to generate the predictors so sorely needed for the intelligent operation of a unit for the care of the acutely ill patient.

VII. A PRACTICAL COMPREHENSIVE PATIENT-MONITORING SYSTEM

A general scheme for comprehensive patient monitoring with realizable hardware and software requirements is shown in Fig. 10-2.

In this scheme primary data items are derived from four sources: (1) from analog data obtained by the inbed electronics per se, requiring

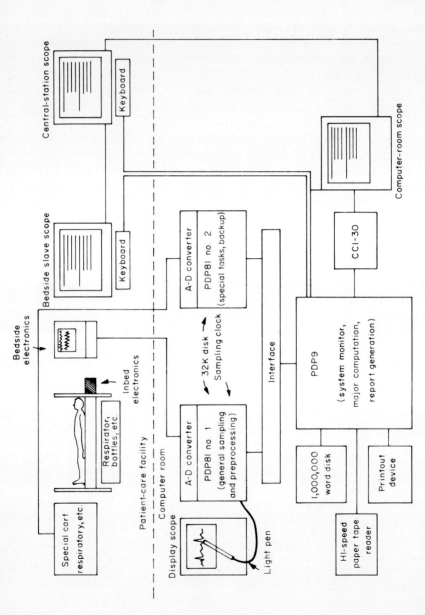

FIG. 10-2. A diagram of the system assembled at the University of North Carolina for comprehensive patient monitoring in the intensive-care unit.

10. Comprehensive Patient-Monitoring

analog-to-digital conversion for interpretation by the computer, (2) from derived information obtained from the bedside electronics, including such items as the heart rate, systolic and diastolic pressure, PVC count, and any alarms generated at the analog level (used to activate computer interrupt circuits), (3) from analog data obtained from special apparatus carts, such as the respiratory-analysis cart shown in the diagram, and (4) by manual entry of data obtained from the clinical laboratory, entry of physician and nurse notes, etc.

The highest sampling rates encountered in this scheme are those required for the sampling of ECG signals (200/sec). It is significant that with the PVC counter operating, it is necessary to carry out ECG analysis only about once every 30 minutes, unless such analysis is indicated between regular sampling intervals. In any case the analysis in this scheme is carried out by using oscilloscope display of only a few complexes, with light-pen selection of significant points on the curves. Data transmission to the large computer then amounts to only a very few data words at a time, requiring only microseconds. The computer analysis of these data is a simple task.

Other sampling rates are quite slow, the fastest being that for respiratory analysis, about 40/sec. Furthermore, these analyses are done seldom enough that they can be carried out with ease on a simple time-sharing basis, so that the hardware system can be used for monitoring a number of patients.

One of the most significant of the hardware items in this system is the existence of three computers rather than one large computer. This assembly provides redundancy, thus failure-proofing the system to a large extent. Also important is the existence of at least one mass-storage device, in this case a disk holding 1 million 18-bit words. Interfacing the computers, using data-break facilities, has not been difficult.

Another important item in this scheme is the CCI-30 data-display device, which is used for the display of all reports generated by the computer and for reactive communication between the computer and the medical staff. This unit consists primarily of a memory capable of holding 800 coded characters (i.e., a character matrix of 40 characters per line and 20 lines per display). The unit displays the characters stored in its memory without tying up the computer. As manufactured this unit has the capability for two-way communication, with a keyboard working through the unit to provide the reactivity desired. In our scheme we have chosen to use the device as a one-way communications path—computer to display only. The reactive keyboards located at each at bed unit and at the central nurse's console are cabled directly to the computer. The advantages of this are (a) that the interfacing of the CCI-30 unit with the computer for one-way communication was much easier and more reliable than for two-way communication, and (b) it was easier to fit the keyboard responses into the overall computer monitor program when the keyboards were connected directly to data-input buses.

With the word length of the PDP-9 computer it is possible to store two characters per data word. Therefore, one entire report matrix occupies only 400 words of disk space. With a million-word disk it is easy to see that it would be possible to store and have available as many as 500 such report formats, using up only 20% of the disk space. By judicious placing of report formats on the disk it is possible to minimize disk search time and bring out the reports, fill in the blanks from table look-up, and transfer the fully generated report to the CCI-30 unit in a few milliseconds.

This system is not appreciably more complex than many of those already set up and operating on a much less comprehensive basis. At least part of the simplification has been accomplished through elimination of what in our research team's opinion have been uselessly frequent measurements and the elimination of computer-based processing where analog processing or human intervention could simplify the entire process.

We should in all fairness point out that at the time of this writing the system described herein is only now in the evaluation phase, and will not be turned over to the clinical facilities until we are satisfied that the system is working at all communications levels. So far, this is a research effort; it is described here only to point the way to thinking about patient monitoring in the intensive-care situation.

VIII. SUMMARY

This chapter has set out to present and defend the thesis that patient monitoring should be considered from the broad point of view that the ultimate goal of such monitoring should be the functional assessment of the critical physiological systems. Such assessment, when properly quantitated and ultimately analyzed using existing systems-study techniques, can result in the determination of predictors of the probability of the patient's entry into crisis, and in crisis may be used to assess the probability of death. More important, perhaps, the information required to obtain such predictors would surely reveal the nature of the deviant condition and allow (in many cases) for corrective measures to be taken before a condition of crisis occurs. Involved in this entire scheme of measurement and analysis is the concept that such studies would also tell us a great deal about the nature of the processes of death.

We also hold in this discussion that in order to carry out the goals of this kind of patient monitoring, the monitoring must be more comprehensive than that carried out in any of the existing units, even those that have been computerized to a large extent. Comprehensive monitoring can be done with methods of measurement and analysis that already exist. The process is aided by the minimizing of sampling-rate requirements, storage requirements, and computing-time requirements by making maximal use of analog preprocessing of physiological signals and of the unique human capabilities, particularly in the area of pattern recognition. There is little question that eventually systems will evolve in which the entire process can be successfully and practically automated; at the moment it would seem judicious to carry on the monitoring with semiautomation.

Finally, a practical system for comprehensive patient monitoring is presented, using what the authors believe to be the most practical methods for acquisition of data, preprocessing of those data, provision of redundancy, provision for reactive communication between computer and medical staff, production of both temporary and hard-copy reports, and bulk storage used to maximum advantage for operation of the system.

References[6]

Blackwood, W. D. (1965). Some practical aspects of the measurement of acid-base balance in blood. *Arch. Inter. Med.* **116**, 654.

Cacares, C. A., C. A. Steinberg, S. Abraham, W. J. Carbery, J. M. McBride, W. E. Tolles, and A. E. Rikli. (1962). Computer extraction of electrocardiographic parameters. *Circulation*, **25**, 356.

Coulter, N. A., Jr. (1968). Toward a theory of teleogenetic control systems. *Gen. Syst. Yearbook*, **13**, 85.

Coulter, N. A., Jr., and O. L. Upkike, Jr. (1965). Biomedical control developments. *Proc. Joint Automatic Control Conf.*

Damman, J. F., D. J. Wright, O. L. Updike, J. D. Grandine, and D. R. Ryan. (1964). Data acquisition and interpretation system for postoperative patients. *Proc. San Diego Symp. Biomed. Eng.*, p. 253.

Feezor, M., and R. W. Stacy. (1969). An analogue electronic detector for premature ventricular contractions. In preparation.

Grodins, F. S. (1963). "Control Theory and Biological Systems." Columbia Univ. Press, New York.

Lewis, F. J. (1967). Computer techniques in the care of surgical patients. *Surgery*, **62**, 630.

Lewis, F. J., and J. E. Jacobs. (1967). Postoperative monitoring. *IEEE Conv. Record*.

Lewis, F. J., T. Shimizu, A. L. Scofield, and P. S. Rosi. (1966). Analysis of respiration by an on-line digital computer system; clinical data following thoracoabdominal surgery. *Ann. Surg.* **164**, 547.

Lown, B., A. M. Fakhro, W. B. Hood, Jr., and G. W. Thorn. (1967). The coronary care unit; new perspectives and directions. *J. Am. Med. Assoc.* **199**, 156.

---

[6] See also references for Chapters 8 and 9.

Osborn, J. J., F. J. Segser, S. E. Elliott, and F. Gerbode (1968). Oxygen uptake and pulmonary function in critically ill patients by analogue computation. *Med. Res. Eng.*

Vallbona, C., W. A. Spencer, L. A. Geddes, W. F. Blose, and J. Canzoneri, III. (1966). Experience with on-line monitoring in critical illness. *IEEE Spectrum*, p. 136.

Warner, H. R., R. M. Gardner, and A. F. Toronto. (1968). Computer-based monitoring of cardiovascular functions in postoperative patients. Circulation, Supp. II, **38**, 68.

Weil, M. H., H. Shubin, and W. Rand. (1966). Experience with a digital computer for study and improved management of the critically ill. *J. Am. Med. Assn.* **198**, 147.

Welkowitz, W. (1964). Postoperative monitoring system. *Ann. N. Y. Acad. Sci.* **118**, 400.

Author Index

Numbers in italics refer to pages on which the complete reference is listed.

Abraham, S., *275*
Amento, J. S., 19, *53*
Armimizu, M., 146, *158*
Armstrong, W., *237*

Backus, J. W., 204, *206*
Beaumont, J. O., 209, 213, 230, *237*
Bender, M. A., 146, *158*
Biggs, H. A., 19, *53*
Blackwood, W. D., *275*
Blaedel, W. J., 26, *52*
Blaivas, M. A., 19, *52*
Blau, M., 146, *158*
Blose, W. F., *276*
Boddington, M. M., 112, *143*
Bonner, R., 127, *143*
Bostrom, R. C., 110, 136, *143*
Boyse, E. A., 141, *144*
Brecher, G., 19, *53*
Brewer, D., 19, *53*
Brown, R. D., 137, *144*
Buckthal, Paul E., 65, *85*
Bunting, S. L., 19, *53*
Burchell, H. B., 146, *158*

Cacares, C. A., *275*
Canzoneri, J., III, *276*
Carbery, W. J., *275*
Carl, M., 141, *144*
Carr, W. F., 16, *53*
Caspersson, O., 141, *143*
Caspersson, T., 109, 136, *143*
Celada, F., 141, *143*
Chaapel, D. W., 146, 149, 157, *158*
Clark, M. A., 24, *53*

Clark, W. A., 22, *53*, 163, *178*
Cole, R. R., 26, *53*
Constandse, W. J., 19, *53*
Cornwall, J, B., 110, *143*
Cotlove, E., 19, *53*
Coulter, N. A., Jr., 255, 257, *275*
Coulter, W. H., 110, *143*
Cox, J. R., 186, 189, 191, 204, *206*
Cunningham, J. R., 165, 168, 175, *178*, *179*

Damman, J. F., 208, *237*, *275*
Davison, R. M., 110, *143*
Day, H. W., 182, *206*
Deal, C., 217, *237*
Del Guercio, L., 208, *237*
Derman, H., 110, 111, 136, 137, *144*
Deutch, A. D., 142, *143*
Diamond, R. A., 112, *143*
Dimsdale, B., 104, *105*
Dolan, C. T., 146, *158*
Dow. P., 224, *237*

Eberhart, R. C., 209, 223, *237*
Elliott, S. E., 217, 221, *237*, *276*
Ellis, E., 217, *237*
Evenson, M. A., 45, 47, *53*

Fakhro, A. M., 182, *206*, *275*
Feezor, M., 270, *275*
Fellows, G. E., 44, *53*
Ferrigan, M., 182, *206*
Flynn, F. V., 19, *53*
Fozzard, H. A., 189, 191, *206*
Fulwyler, M. J., 111, *143*

AUTHOR INDEX

Gardner, R. M., 240, 244, *251*, *276*
Geddes, L. A., *276*
Gelerenter, H. L., 113, *143*
Gennaro, W. D., 19, *53*
Gerbode, F., 217, 213, 221, 230, *237*, *276*
Gianelly, R., 182, *206*
Gieschen, M. M., 19, 44, *53*
Giesler, P. H., 18, *53*
Ginsburg, S., 204, *206*
Grandine, J. D., *275*
Grodins, F. S., 255, *275*
Gupta, S. K., 168, *178*

Habig, R. L., 45, *53*
Hamilton, W. F., 224, *237*
Harrison, D. C., 182, *206*
Hicks, G. P., 16, 19, 26, 44, 45, *52*, *53*
Highleyman, W. H., 113, 127, *143*
Hilberg, A. W., 110, *144*
Hood, W. B., Jr., 182, *206*, *275*
Hoyt, R. S., 19, *53*

Ingraham, S. C., 110, *144*

Jacobs, J. E., *275*
Jason, D., 221, *237*
Johns, H. E., 163, 165, 168, 175, *179*

Kaiser, R. F., 110, *144*
Kakehi, H., 146, *158*
Kamentsky, L. A., 110, 111, 113, 118, 122, 127, 131, 136, 137, 141, *143*, *144*
Keane, J. F., Jr., 136, *144*
Keenan, J. A., 45, 47, *53*
Kinsman, J. N., 224, *237*
Kirkham, W. R., 19, 44, *53*
Kirsch, W. J., 26, *53*
Kitchell, J. R., 182, *206*
Klionsky, B., 19, *53*
Koenig, S. H., 137, *144*

Ladinsky, J. L., 136, *144*
Laessig, R. H., 26, *52*
Lambrinidis, Philippos, 65, *85*
Lamson, B. G., 18, *53*, *104*, *105*

Larson, F. C., 19, 44, 45, *53*
Lewall, D. B., 148, *158*
Lewis, F. J., 208, *237*, *275*
Lindberg, D. A. B., 19, *53*
Liu, C. N., 118, 127, *144*
Lomakka, G., 109, *143*
Lown, B., 182, *206*, *275*

McBride, J. M., *275*
McClung, D., *237*
Mazzia, V. D. B., 221, *237*
Melamed, M. R., 110, 111, 122, 131, 136, 137, 141, *144*
Mellors, R. C., 136, *144*
Meltzer, L. E., 182, *206*
Mendelsohn, M. L., 118, *144*
Messing, A. M., 141, *144*
Miller, C. W., *237*
Moldavan, A., *144*
Molnar, C. E., 22, *53*, *163*, *178*
Montgomery, P. O'B., 112, *144*
Moore, J. W., 224, *237*
Mueller, William, J., 65, *85*
Murphy, T. W., 221, *237*
Murray, N. A., 136, *144*

Nolle, F. M., 189, 191, *206*
Noto, T., 19, *53*

O'Brien, R. T., 142, *144*
Oliver, G. C., Jr., 189, 191, *206*
Ornstein, L., 122, *144*
Osborn, J. J., 209, 213, 217, 230, *237*, *276*

Palmon, F., 182, *206*
Papanicolaou, G. N., 136, *144*
Pateau, K., 122, *144*
Peacock, A. C., 19, *53*
Peckham, B. M., 16, *53*, 136, *144*
Pekores, J., 182, *206*
Peters, R. M., 221, *237*
Piper, K. A., 19, *53*
Potter, R. J., 110, 113, *143*, *144*
Prewitt, M. S., 118, *144*
Pribor, H. C., 19, 44, *53*

Pruitt, J. C., 110, *144*
Pryor, T. A., 240, 241, 244, *251*

Raison, J. C. A., 209, 213, 230, *237*
Rand, W., *276*
Rappaport, A. E., 19, *53*
Reed, C. E., 16, *53*
Rikli, A. E., *275*
Ritter, W., 141, *144*
Roberts, P. K., 19, *53*
Rotman, B., 141, *143*
Rosenthal, E., 141, *144*
Rosi, P. S., 208, *237, 275*
Russell, J. A. G., 209, 213, 230, *237*
Ryan, D. R., *275*

Sandritter, W., 141, *144*
Sarto, G. E., 136, *144*
Sawyer, H. S., 110, 136, *143*
Schlein, B. M., 45, *53*
Schoop, R. A., 19, *53*
Schultz, Karl E., 65, *85*
Scofield, A. L., 208, 237, *275*
Sebestyen, G. S., 127, *144*
Sedlis, A., 137, *144*
Segger, F. J., 217, 221, *237, 276*
Shimizu, R., 208, *237*
Shimizu, T., *275*
Shubin, H., 208, *237, 276*
Siegel, J. H., 208, *237*
Silver, R., 136, *144*
Slack, W. V., 16, 19, 22, 44, *53*
Smith, S. J., II, 110, *144*
Souer, H., 182, *206*
Spencer, W. A., *276*
Spivack, A. P., 182, *206*
Spraberry, M. N., 19, *53*
Sprau, A. C., 146, 149, 157, *158*

Spurling, R. G., 224, *237*
Stacy, R. W., 221, *237*, 270, *275*
Stauffer, W. M., 240, 244, *251*
Steinberg, C. A., *275*
Struamfjord, J. V., 19, *53*
Svensson, G., 109, *143*
Sweet, R. G., 111, *144*

Tauxe, W. N., 146, 148, 149, 150, 152, 175, *158*
Thiers, R. E., 26, 45, 47, *53*
Thorn, G. W., 182, *206, 275*
Tolles, W. E., 110, 136, *143, 275*
Toronto, A. F., 240, 244, *251, 276*

Uchiyama, G., 146, *158*
Unger, S. H., 112, *144*
Updike, O. L., Jr., 208, *237*, 255, *275*

Vallbona, C., *276*
Van Cura, L. J., 16, 22, *53*
von der Groeben, J. O., 182, *206*

Wald, A., 221, *237*
Walsh, Leo F., 65, *85*
Walters, L., 45, *53*
Warner, H. R., 208, 240, *237*, 244, 251, *276*
Wattenburgh, W. H., 19, *53*
Weid, G. L., 141, *144*
Weil, M. H., 208, *237, 276*
Welkowitz, W., 208, *237, 276*
Williams, G. Z., 16, 18, 20, *53*
Willoughby, M. B., 110, *144*
Wright, D. J., 208, *237, 275*

Subject Index

A

Absorption, optical, *see also* Spectrophotometry
  Beer's law in, 120
  by blood cells, 108, 136
  by cancer cells, 121, 136
  in cell identification, 119
  Lambert's law in, 119
  of lymphocytes, 142
  multiple-wavelength, 137
  of nucleic acids, 136
Acid-base balance, in intensive care monitoring, 213
Actuarial analysis, *see* Statistics
Alarms, in intensive care monitoring, 229
Analog-to-digital conversion
  in cell scanning, 134
  in clinical laboratory, 21, 27
  in ECG processing, 190, 201
  in patient monitoring, 265
Arrhythmias, cardiac, *see* Heart
ASCII code, use in storing patient data, 64
Atrial fibrillation, 199
AutoAnalyzer
  automation of, 25
  computer control of, 59
  record of, 56
  interfacing of, with computer, 66
  performance of, computerized, 47
  sample separation in, 26
Automation, of laboratory procedures, 17
Averaging, in clinical laboratory work, 73
AZTEC (Amplitude Zone Time Epoch Coding), 190 ff
  evaluation of, 200

B

BASP (Biomedical Analog Signal Processor), 60
  cost of, 83
  flow chart for programming of, 73
Beam data (radiation treatment planning), 166
Beam generator program for radiation treatment planning, 168
Beer's law, in optical absorption by cells, 120
Berkson and Gage algorithm, 95
Biopsy, use to standardize cancer cell screening, 137
Blood, *see also* Leukocytes
  cells,
    differential counting of, 108, 136, 142

C

California Tumor Registry, 96, 103
Cancer
  cervical, 138, 139
  detection of, 136
    by computers, 108
  epidermoid, 137, 138, 140
  identification of, 88
  optical absorption by, 121, 136
  screening for, 109, 110, 136
Capacity, of LINC computer, limitations imposed by, 48
Cardiac output, *see also* Heart
  determination of, 223 ff
  in patient monitoring, 243
Carnoy fixative, as cell suspension medium, 137

SUBJECT INDEX

Cascaded processors, *see* Processor, cascaded
Catastrophic failure, as concept of death, 255
Catheterization, for intensive care monitoring, 243
CCI-30 display device, 273
CDC 16a, *see* Computers
Cell counting, 2
Cells
　cancer, photographed at 2600A, 138, 139
　characterization of, 109
　classification of, 123, 126
　differentiation of, 136
　identification of
　　automatic, 107 ff
　　by fluorescence, 123
　　by optical absorption, 119
　measurement of, automatic, 117
　quantized images of, 114, 115
　recognition of, 113
　scanning of, 112
　sorting of, 107 ff
　　for automatic identification, 111
　with rapid cell spectrophotometer, 133
　staining of, 117
　transport of, in cell identification, 109
　　liquid, 110
　　on tape, 112
　treatment of for automatic sorting, 111
　tumor, 131
　vaginal, 132
　　histogram classification of, 123-125
Character generator, in display units, 64
Chromosome karyotyping, 127
Clinical areas, automation of,
　areas amenable to, 58
　problems with, 55
Clinical chemistry, 7
Clinical laboratory, automation of
　actual experiences, 45
　evaluation of, 51
　system critique, 50
　　computer use in, 70
　　problems of, 56, 57
Cobalt-60 beam calculation, 173
Cognition, simulation of, 11, 12
Communication, computer-physician, in intensive care monitoring, 211, 242

Compliance, lung, 217, *see also* Respiration
Comprehensive patient monitoring, *see* Monitoring, patient
Computer
　design for data transformation, 186
　in intensive care monitoring, 209
　　design for, 214
　small, in radiation treatment planning, 175
　specialized, for radiation treatment planning, 160
Computer failure, 77
　experience with, 51
　redundancy prevention, 243
Computer-physician communication, in patient monitoring, 271
Computer time, in multiaccess system, 75
Computers
　CDC 160A, 57
　CDC 3200, 241
　clinical uses of, 58, 59
　hardware requirements for clinical laboratory, 21
　IBM 360/40, 96
　IBM 360/75, 96
　IBM 1130, 130, 145
　IBM 1800, 209
　IBM 7040, 147
　requirements for, clinical, 59
　satellite, 61, 69, 83
　SEL 810A, 61
Context
　relative to ECG data stream, 187
Contrast enhancement, of scintiscan results, 146
Control file in tumor registry, 97
Control systems, 257
　changes in illness and death, 257 ff
Control Data 3200 computer, 241
Conversational mode, *see* Reactive communication
Coronary care unit, 182, *see also* Intensive care unit
Costs
　of alarms generation in ICU, 230
　of computation in hospital system, 83, 85
　of computerizing clinical laboratory, 19
　of computers for clinical use, 59
　of hospital service, 79

Costs (*cont.*)
  of multiaccess computer systems, 74, 75
  of patient care, computerized, 2
Creatinine, clearance of, calculation by LINC computer, 43

D

Data
  acquisition of, in clinical laboratory, 21, 27
  amounts of, in patient monitoring, 265
  manipulation of, in radiation treatment planning, 169
  rates, in ECG processing, 201
  storage of
    in clinical data registries, 88
    in clinical laboratory, 20, 21
    mass, 9
    of patient data, 51, 63
    problems of, 87 ff
    in tumor registry, 96
  transformation of
    for cell sorting, 127
    in electrocardiography, 181 ff
    in scintillation scanning, 145 ff
Data master in programmed console, 163
Death
  nature of, 254 ff
  probability of, 259, 260
  relation to monitoring, 254
DEC PDP-9 computer, 274
Delay line, in computer system design, 61
Diagnosis, hemodynamic, 232
Diagnostics, in quality control for clinical laboratory, 46
Differential blood cell count, 142
Digitization, *see also* Analog-to-digital conversion
  of radiation dose levels, 167
Disk storage
  data transfer time cost of, 60
  in hospital system, 83
  and IBM 1130, 134
  on LINC computer, 49
  in multiaccess system, 68, 75, 76
  in patient monitoring, 274
  tumor registry storage and, 97, 103
Diseases, classification of, 77

Display, consoles, 62
  via FORTRAN, 99
  graphic, 3
  in hospital systems, 78
  IEE numerical, 65
  in LABMON, 33
  in maintenance program of clinical lab monitor, 41
  in multiaccess system, 75, 76
  on-line, 87 ff
  patient data retrieval and, 88
  in patient monitoring, 210–212, 242, 246, 248, 273
  in radiation treatment, 159 ff
    planning, 170
  in tumor registry, 89–92, 96
Distance functions in cell sorting, 126
DNA, *see* Nucleic acids
Dosage distribution in radiation treatment planning example, 177
Double blind tests in cell scanning, 137

E

Electrical safety, *see* Shock hazard
Electrocardiography, 181 ff
  processors for, 188–190
  sampling rates in, 183
    block diagram of, 189
Electrocardiogram
  changes in, 183
  classification of, 183
  data storage of, 186–188
  in intensive care monitoring, 227 ff
  on-line monitoring of, 2
  parameter histograms of, 200–202
  processing procedures for, 184
  processor for
    atrial channel, 196
    cycle processor, 197
    program details, 198
    ventricular channel, 190
  reconstruction of, 196
  transformation of
    evaluation of, 200
    program for, 198
    theory of (special appendix), 205
Enter doses program (radiation treatment planning), 167, 168

SUBJECT INDEX

Error, in automation of clinical laboratory, 45
Executive system, use in intensive care monitoring, 216
Exercise, test plots of, 235
Exercise laboratory, in intensive care system, 233

F

Fan lines (radiation treatment planning), 167
File structures, see Data, storage
Filter, analog, in ECG processing, 190
Flow, in cell scanning, 130, 131
Flow, respiratory, errors of measurement of, 221
Fluid balance, in intensive care monitoring, 247
Fluid switches, in cell sorting, 111
Fluorescein diacetate, in cell viability assay, 141
Fluorescence, in cell identification, 118, 123
FORTRAN
 in multiaccess system, 73, 74
 in tumor registry search, 99, 103

G

Grade subprocessor, see AZTEC
GRAF (Graphic Addition to Fortran), 99
Graphic display, in tumor registry search, 102
Graphing machine (for distribution of cells on transport tape), 112
Gray tones, in scintiscan interpretation, 145
Grounding, and shock hazard, 216, 217

H

Heart
 arrhythmias of, 182
 cycles of, description by AZTEC, 198
 disease of, photoscanning in, 149
 failure of, 261
 monitoring of, 245 ff
 rate, 227
 stroke volume of, 248

Hemodynamics, in intensive care monitoring, 232
Hemothorax, detection in intensive care monitoring, 236
Histograms
 of ECG data from transformations, 200
 generation of by LINC computer, 44
 of human blood cells, 128
 of vaginal cells, 124, 125
Histology codes, 96
Holistic approach in clinical laboratory, 19
Hollerith code, and patient data entry, 65
Hospital, computation of, 16, see also Hospital systems
Hospital number, see Patient data
Hospital systems
 business data handling in, 81
 general procedures of, 80

I

IBM 1130 computer
 in cell scanning, 130
 photograph of, 135
IBM 1800 computer, in intensive care monitoring, 209
IBM 7040 computer, in scintiscan analysis, 147
Identification
 of cell scans, 135
 patient, 147
 in scintiscan analysis, 147
Image conversion, see Pattern recognition and scanning
Image processing, 3, 11
Inflections (in ECG processing), 194
Information content, of cell images, 108
Input/output, in tumor registry search, 103
Instruction (computer assisted), 12
Instrumentation, for intensive care monitoring, 209 ff, 225, 244, 268, 269
Intensive care, 207 ff, see also Coronary care unit
Intensive care monitoring, environmental limitations of, 229
Interaction matrices, in biological control systems, 259

Interface, control of rapid cell spectrophotometer, 134
International statistical classification of neoplasms, 96
Interrupts
  use in multiaccess system, 69
  tumor registry search, 102
Isodose charts, computation of, 164, 165

K

Keyboard, for data entry, 82
Keypunch, use with rapid cell spectrophotometer, 136

L

LABCOM, 16
LABMON, 28
  computer time required for, 29
  diagnostics in, 46
  flow diagram for, 31
  function of, 27
  memory allocation for, 28
  performance of, 34
  procedural sequence of, 35
  programs of, 29
Laboratory, clinical, 15 ff
  computer use in, 19
Laboratory specimens, handling in clinical laboratory system, 18
Lambert's law, in optical absorption by cells, 119
Languages, computer, 6
LAP (assembly language for LINC computer), 24
Latter Day Saints Hospital, 239, 269
Leukocytes, computer recognition of, 118
Light pen
  use in inquiry procedures, 88
  tumor registry search, 103
LINC computer, 163
  capacity of, 51
  in clinical laboratory, 16, 42
  as desk calculator, 44
  expansion of, 50, 52
  general description of, 22
  interface with other devices, 24, 26
  limitations of, 51
  physical examination, 16
  programming, on-line use, 30
  as reactive device, 16, 27, 51
LINC-8 computer system (illustration), 23
List-processing by computers, 9, *see also* Data storage
Liver, scanning of, 153, 155
Lung, photoscans of, 150
Lung mechanics, *see* Respiration
Lymphocytes, rapid cell spectrophotometer scan of, 142

M

Magnetic card, in data master, 163
Mass constituent representation in cell identification, 118
Mathematical modeling, 7, 10, 11
Matrix, data, for scintiscan, 146 ff
Medical electronics, *see* Instrumentation
MEDLAB (patient monitoring system), 241 ff
Microscopy
  in cell scanning, 121 ff
  dark-field in viability studies, 141
Miss rate, in cell scanning, 141
Mitral stenosis, chest photoscan of, 149
Model building, *see* Mathematical modeling
Monitor (software), for clinical laboratory system, 28
Monitoring, coronary, analog, 182
Monitoring
  intensive care, 207 ff
    aims of, 208, 240
    alarms in, 229
    hemodynamics in, 232
    nursing staff and, 236
    parameters measured, 213
    scheme of, 243
    system evaluation, 234
  patient, 239 ff
    aims of, 263, 264
    BASP capability, 84
    calculable quantities in, 266
    comprehensive, 253 ff
    critical systems in, 264
    on-line, 58

philosophy of, 7
requirements for, 265
as research procedure, 263
state of art, 268
Multiaccess system, 55
clinical experience with, 79
critique of, 79
evaluation of, 82
operation of, 69 ff
performance of, 74 ff
Multidisplay (program in radiation treatment planning), 171
Multifont print readers, 118
Multiplasic screening, *see* Screening
Multiple wavelength spectrophotometry, in cell scanning, 137
Multiplexer, use in cell scanning, 134
Multiplexing, of I/O equipment, 61
Multivariable analysis, 259
Myocardial infarction, 183, *see also* Heart

N

National Center for Health Services Research and Development, 4
Noise
as factor in cell scanning, 112
reduction by mechanical damping, 25, 26
Normality, quantitative definition of, 256
Nucleic acids
in cancer cell photometry, 136
detection, 141
Nurse activity scheduling in patient monitoring, 249
Nurse's notes in patient monitoring, 249, 265

O

On-line measurement, in intensive care, 207 ff
Orders, physicians, handling of, 56
Organism system theory, 258, 259
Ornstein and Pateau 2-wavelength method for cell identification, 122
Output
of cycle processor (ECG), 199
of ECG preprocessors, 195

graphic
in intensive care monitoring, 213
of programmed console, 161
in radiation treatment planning, 168
hardcopy, 171
Oxygen uptake, calculation of, in intensive care monitoring, 218

P

Papanicolaou technique, as standard for cancer cell screening, 137
Partitioning
hyperplane, in cell classification, 126
hypervolume, 126
Patient contour program in radiation treatment planning, 166
Patient data, entry of, 70, 80
Patient file, *see* Data storage
Patient identification, entry in patient monitoring, 63, 71, 244, 245
Patient monitoring, *see* Monitoring
Patient records, *see* Records
Pattern recognition, 12
algorithms for, 112, 113
of cells, 108, 117
Phantom, use in radioisotope scanning, 156, 157
Photomultipliers, in cell scanning, 134
Photoscan, 146 ff, *see also* Scintiscan
Physician orders, in patient monitoring, 265
Physicians, response to tumor registry search program, 104
Plotter
use in cell scanning, 133
intensive care monitoring, 213
scintiscan, 147
Pointillistic mode of scintiscan presentation, 145
Predictors, of crisis and death, 260
Premature atrial contraction (PAC), 199
Premature ventricular contraction (PVC), 199
analog detection of, 269, 270
Preprocessing, of AutoAnalyzer signals, 66, 67
Pressure-volume plots in respiration, 219
Primitive processor (as part of ECG processing), 191

Primitive processor (*cont.*)
  diagram of, 192
Printer
  for AutoAnalyzer preprocessor, 68
  for hospital system, 82
Printout
  from LINC computer in clinical laboratory monitoring, 49
  from multiaccess system, 72
  of scope display (tumor registry), 95
Probability theory, in prediction of crisis and death, 260 ff
Process control in medicine, 3
Processor, cascaded
  advantages of, in electrocardiography, 184, 185, 187
  system for ECG analysis, 188–190
Programmed console, 161
  input devices for, 162
Programs
  maintenance, in clinical laboratory monitoring, 40
  user controlled, in clinical laboratory, 33
Pulse height analyzer, in scintiscan analysis, 146, 147
P wave (of ECG), identification of, 197

Q

Quality control, in real-time applications, 45

R

R wave, detection of, 228
R wave (of ECG), identification of, 194
Radiation, dosage distribution, calculation of, 160
Radiation dosage profile, 174
Radiation dosimetry, 2
Radiation therapy, planning of, 163
Radiation treatment planning, 159 ff
  errors of, 172
  examples of, 175–177
  evaluation of, 171, 173
Radioisotopes, organ uptake of, 148
Radioisotope scan processing, 145 ff
Rapid cell spectrophotometer (RCS), 127 ff
  clinical applications of, 136 ff
  construction of, 127–132
  diagram of, 130
  total scheme diagram, 135
Reactive communication
  in automatizing clinical laboratory, 21
  with computers, 5
  in LABMON, 34, 35
Reactive programming, in tumor registry search, 102
Readout
  of scanning procedures, 156
  of scintiscan data, 147
Real-time applications, quality control in, 45
Records, patient, from clinical laboratory, 18
Redundancy
  in electrocardiography, 183 ff
  in patient monitoring, 273
  system reliability and, 78
Reliability
  of computer systems, 77
  requirements in tumor registry search, 104, 105
Remote terminals, 63, 64
  in clinical laboratory systems, 50, 51
  in tumor registry search, 104
Reports
  from clinical laboratory, 18
  in clinical laboratory monitoring, 38, 39
  generation of, 37
  in LABMON, 37
  in patient monitoring, 270
Research data, computer processing of, 59
Research in patient care environment, 264
Respiration
  analysis in intensive care monitoring, 217 ff
  failure of, 261
  measurement in intensive care, 213
  parameters of, 222, 223
Respirator
  use in intensive care, 217, 218
  malfunction of, 213
Retrieval, *see* Data storage
Rho-theta device
  as computer input, 161
  programming for, 168
RNA, *see* Nucleic acids
Rose Bengal, in liver scanning, 153, 155

SUBJECT INDEX

S

Sampling rates, in patient monitoring, 273
Scanner, modulated spot, 118
Scanning
 of cells, 112, 113
 of microscope fields, 109
 radioisotope, 3
 signal-to-noise ratio of, 112
Scatter, of light, in cancer cell screening, 137
Scintiscan, 145 ff
 computer processing of, 158
 reprocessing of, 148
 technique of, 146 ff
Screening
 for cancer, 109, 110
 multiphasic, 8
 of patient data (monitoring), 240
SEL 810A computer, 61
Service requests, in multiaccess system, 71
Shock hazard, in intensive care monitoring, 216
Shock, patient monitoring in, 7, 8
Signal-to-noise ratio, in cell identification, 117
Slides, microscopic, in cell identification, 109
Snowballing, in death processes, 261
Software
 for clinical laboratory data acquisition, 27
 for medical computing, 6
 for multiaccess system, 69
 on line, LINC computer, 30
 for tumor registry search, 98, 100, 101
Spectrophotometry in cell sorting, 126, 127
Spectrophotometer
 cell, 119
 rapid cell, *see* Rapid cell spectrophotometer
SSD method for radiation dosage correction, 175
State variable analysis, 260
Statistics
 actuarial analysis in tumor registry, 94
 of tumor registry, 93
Stroke volume, *see* Heart
System concept, in clinical laboratory, 20

Systems, biological, in life and death processes, 255
Staining of cells for identification, 110
Standards, use in clinical laboratory, 72, 73
Statistics, in LINC computer, 44
Superimposition of beams program in radiation treatment planning, 169

T

Tape transport, of cells, 112
Terminals, interactive, 8
Thyroid
 phantom of, in scanning, 157
 scintiscan of, 151, 152
Time sharing, and patient data retrieval, 88
Transfer functions, *see* Control systems
Trypan blue, in cell viability assay, 141
Tumor
 brain (scanning of), 153, 154
 detection of by scanning, 152
 radiation treatment of, 175
Tumor nomenclature, American Cancer Society manual of, 96
Tumor registry, 87 ff
 critique of, 103 ff
 flow diagram (search), 100
 performance of, 102 ff
 system design for, 104
 uses of, 103

U

User control, of programs in LABMON, 32

V

Vaginal cells, *see* Cells, vaginal
Ventricular ectopic beats, 227, *see also* Premature ventricular contraction (PVC)
Ventricular subprocessor, *see* AZTEC (Amplitude Zone Time Epoch Coding)
Viability
 of cells
  assay of, 108, 136, 141
 of whole organism, 258

W

Washington University Biomedical Computer Laboratory, 160
White blood cells, *see* Leukocytes
Wiggle subprocessor, *see* AZTEC (Amplitude Zone Time Epoch Coding)

Work of breathing, 220, *see also* Respiration

Z

Zero resolution scanning of cells, 113